Math Series
ADVANCED
BOOK 2

by Stephen B. Jahnke

Book cover design by Kathy Kifer

Dedicated
to Joy Leal
for her help and insight.

Published by
Garlic Press
100 Hillview Lane #2
Eugene, OR 97401

ISBN 0-931993-38-5
Order Number GP-038

CONTENTS

Graphing Linear Equations

Equations can be better understood by looking at their graphs. A **graph** is a picture that enables us to see how the variables in an equation are related to each other. Graphs of equations are drawn on a grid which is called the **Cartesian Coordinate System**.

The Cartesian Coordinate System (Figure A)

The Cartesian Coordinate System involves two axes (or number lines). The horizontal axis is called the **x-axis** and the vertical axis is called the **y-axis**. The x-axis and the y-axis cross each other at a place called the **origin**. The value of x at the origin is 0. The value of y at the origin is 0. See Figure A.

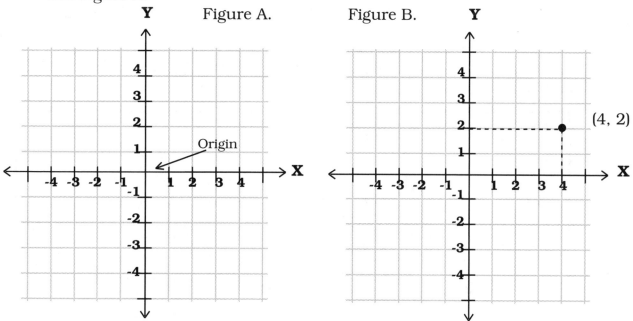

Ordered Pairs (Figure B)

Every point has both x and y values called coordinates. The coordinates can be displayed as an ordered pair. In an **ordered pair**, the x-coordinate is always listed first, and the y-coordinate is always listed second.

For example, in Figure B, the x-coordinate of the point is 4, and the y-coordinate of the point is 2. The point itself is labeled by the ordered pair (4,2), with the x-coordinate listed first.

•Example 1:

Plot each ordered pair:
1. (1,5)   5. (6, -1)
2. (5,1)   6. (0, 3)
3. (-3,2)   7. (-4, -2)
4. (2, 0)   8. (0, -3)

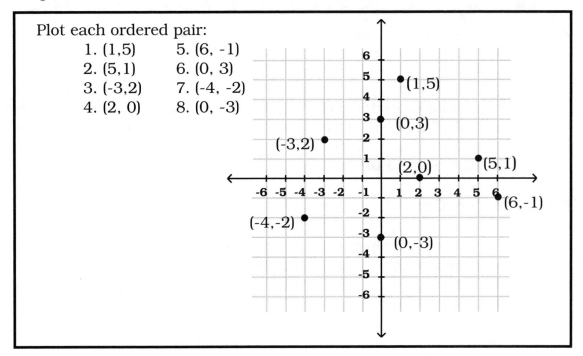

•Example 2:

List the ordered pair for each point.

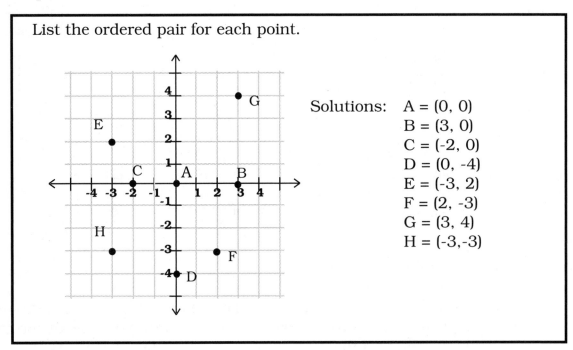

Solutions: A = (0, 0)
             B = (3, 0)
             C = (-2, 0)
             D = (0, -4)
             E = (-3, 2)
             F = (2, -3)
             G = (3, 4)
             H = (-3,-3)

Graphing Linear Equations, Exercise 1.
 For each point below, write the
x and y-coordinates as an ordered pair.

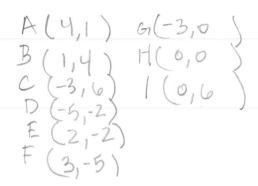

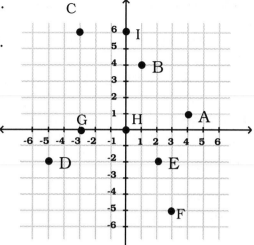

A $(4,1)$ G $(-3,0)$
B $(1,4)$ H $(0,0)$
C $(-3,6)$ I $(0,6)$
D $(-5,-2)$
E $(2,-2)$
F $(3,-5)$

Graphing Linear Equations, Exercise 2.
 Draw your own coordinate system and plot the following:

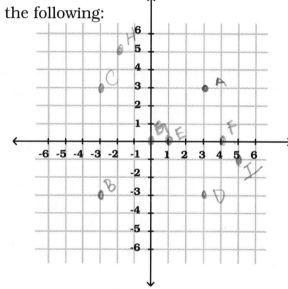

A. (3, 3)
B. (-3, -3)
C. (-3, 3)
D. (3, -3)
E. (0, 1)
F. (4, 0)
G. (0, 0)
H. (-2, 5)
I. (5, -2)

Linear Equations and Solutions

7x - 3y = 12 and y = 2x + 1 are examples of linear equations.

An equation is not linear if the degree of any term is 2 or more. For example, $x^2 = 9$ and $5x^2 + x = 1$ are not linear, since both equations contain terms of degree 2.

The **solutions** of a linear equation are the ordered pairs which make the equation true. For example, (2, 5) is a solution to y = 2x + 1, because replacing x with 2 and y with 5 gives:

        y = 2x + 1
        5 = 2(2) + 1
        5 = 4 + 1
        5 = 5   Which is a true statement.

(0, 1) is also a solution to $y = 2x + 1$, since replacing x with 0 and y with 1 gives:

$$y = 2x + 1$$
$$1 = 2(0) + 1$$
$$1 = 0 + 1 \quad \text{Which is a true statement.}$$

Any replacement of x and y which leads to a false statement is not a solution. For example, the ordered pair (7, 4) is **not** a solution to $y = 2x + 1$:

$$y = 2x + 1$$
$$4 = 2(7) + 1$$
$$4 = 14 + 1$$
$$4 = 15 \quad \text{False.}$$

Graphing Linear Equations, Exercise 3. Do these problems by replacing the variables with the ordered pairs. Remember, the first number in each ordered pair represents the x value.

1. Which of the following ordered pairs are solutions to $y = 3x + 2$?

    a. (0, 2)   c. (2, 1)   e. (-2, 0) $0 = 0 + 2$
    b. (1, 5)   d. (-1, -1)   f. (-3, -7) $5 = 3 + 2$

2. Which of the following are solutions to $x + 3y = 6$?

    a. (0, 2)   c. (-3, 3)   e. (12, -2)
    b. (6, 0)   d. (1, 2)   f. (-2, 1)

3. Which of the following are solutions to $y = -2x - 3$?

    a. (5, -13)   c. (0, -3)   e. (-1, -5)
    b. (4, 0)   d. (-4, -11)   f. (-10, 17)

4. Which of the following are solutions to $4 - x = 2y$?

    a. (0, 2)   c. (2, 1)   e. (-2, 3)
    b. (3, 0)   d. (6, -1)   Hint: $4 - x = 2y$ is the same as $4 - 1 \cdot x = 2y$.

Graphing by Plotting Points

In this section, you will learn the first method for graphing linear equations. **Graphing an equation** means plotting its solutions. The solutions are the ordered pairs that make the equation true. When the solutions of a linear equation are graphed, the result is always a straight line. This is why these equations are called linear (or line) equations.

To Graph a Linear Equation (Method 1)

Find the solutions of the equation by replacing x with three different numbers. The solutions are recorded in a table, then plotted.

•Example 1:

Graph $3x + y = 1$.

    Method: Draw an **x, y-table** to record the ordered pair solutions as we find them.

X	Y

    Find the first solution by choosing any number at random. Let's choose zero. Put 0 in the x-column. They replace x by 0 in the equation and solve for y.

X	Y
0	

$$3x + y = 1$$
$$3 \cdot 0 + y = 1$$
$$0 + y = 1$$
$$y = 1$$

    $y = 1$ means that 1 is the y value which correponds to the x value of 0. Place 1 in the y-column.

X	Y
0	1

    In other words, (0, 1) is a solution to $3x + y = 1$.

To graph the equation, we need two more solutions. Replace x with any two other numbers. Let's try 2 and -1. Enter the numbers in the x column of the table, replace x by those numbers in the equation and solve for y.

X	Y
0	1
2	-5
-1	4

Replacing x by 2.

$3x + y = 1$

$3 \cdot 2 + y = 1$

$6 + y = 1$

$6 + y + (-6) = 1 + (-6)$

$y = -5$

When x is -1.

$3x + y = 1$

$3(-1) + y = 1$

$-3 + y = 1$

$-3 + y + 3 = 1 + 3$

$y = 4$

In other words, the ordered pairs (0, 1), (2, -5), and (-1, 4) are solutions of $3x + y = 1$.

Plot these ordered pairs and draw a line passing through them.

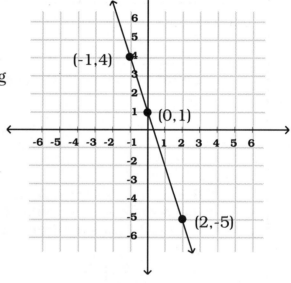

In the above example, we draw a line through (-1, 4), (0, 1), and (2, -5) to indicate that these ordered pairs are not the only solutions. In fact, any other ordered pair which the line passes through is also a solution.

For instance, the above line passes through (1, -2). Thus (1, -2) is a solution to $3x + y = 1$.

To check that (1, -2) is a solution, replace x with 1 and y with -2:

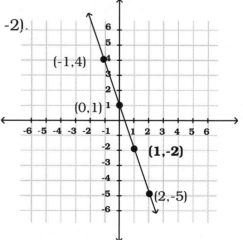

$3(1) + (-2) = 1$

$3 + (-2) = 1$

$1 = 1$

True.

6

Example 1 suggests the following steps:

Step 1: Draw an x, y-table.
Step 2: Pick numbers to replace x and solve for y.
Step 3: Plot ordered pairs and draw a line.

•Example 2:

Graph $2y - x = 4$.

Solution: Replace x with any three numbers. Let's try 0, -4, and 1.

X	Y
0	2
-4	0
1	$\frac{5}{2}$

When x = 0
$2y - x = 4$
$2y - (0) = 4$
$2y - 0 = 4$
$2y = 4$
$y = 2$

When x = -4
$2y - x = 4$
$2y - (-4) = 4$
$2y + 4 = 4$
$2y = 0$
$y = 0$

When x = 1
$2y - x = 4$
$2y - (1) = 4$
$2y - 1 = 4$
$2y = 5$
$y = \frac{5}{2}$

When x = 1, the corresponding y-value is the fraction $\frac{5}{2}$. It may work to use the fraction $\frac{5}{2}$ and plot $(1, \frac{5}{2})$. However, fractions are difficult to plot and sometimes they make the graph of a line inaccurate. Therefore, we will keep trying x-values until we get a third y-value that is not a fraction.

X	Y
0	2
-4	0
1	$\frac{5}{2}$
3	$\frac{7}{2}$
2	3

When x = 3
$2y - x = 4$
$2y - 3 = 4$
$2y = 7$
$y = \frac{7}{2}$ No.

When x = 2
$2y - x = 4$
$2y - 2 = 4$
$2y = 6$
$y = 3$ Yes.

Plot the non fraction solutions (0, 2), (-4, 0) and (2, 3) and draw a line.

•Example 3:

Graph $4 + x = y$.

X	Y
0	4
1	5
-3	1

When x = 0.

$4 + x = y$
$4 + 0 = y$
$4 = y$

When x = 1.

$4 + x = y$
$4 + 1 = y$
$5 = y$

When x = -3.

$4 + x = y$
$4 + (-3) = y$
$1 = y$

Linear Equations, Exercise 4. Graph by making an x, y-table.

1. $2x + y = 3$
2. $y + 5x = 0$
3. $-2x + y = 1$

4. $y = 2x + 1$
5. $y + 4 = 3x$
6. $x - 3 = y$

7. $y + x = -3$
8. $y = 2x$
9. $-x + 1 = y$

Graph (Hint: Keep trying x-values until you get three y-values that are not fractions).

10. $2y + x = 2$
11. $x + 2y = -4$

12. $2y = x + 3$
13. $3y = x + 1$

14. $x = 3y$
15. $-4y = x$

Graphing Using Zeros (Intercepts)

In this section, you will learn the second method for graphing linear equations. This method is similar to the first method in that the variables are replaced by numbers. This time we use the number zero to replace first x, then y.

8

To Graph Linear Equations (Method Two)

Step 1: Replace x with zero and solve for y.
Step 2: Replace y with zero and solve for x.
Step 3: Find a third point by replacing x with any number. This third point is called a **check point**.

•Example 1:

Graph $2x + 3y = 6$.

Step 1.	Step 2.	Check Point. Let's try -3.
When x = 0	When y = 0	x = -3
$2x + 3y = 6$	$2x + 3y = 6$	$2x + 3y = 6$
$2(0) + 3y = 6$	$2x + 3(0) = 6$	$2(-3) + 3y = 6$
$0 + 3y = 6$	$2x + 0 = 6$	$-6 + 3y = 6$
$3y = 6$	$2x = 6$	$3y = 12$
$y = 2$	$x = 3$	$y = 4$

X	Y
0	2
3	0
-3	4

The ordered pairs (0, 2) and (3, 0) are used to graph the line. (-3, 4) is used as a check to make sure nothing has gone wrong.

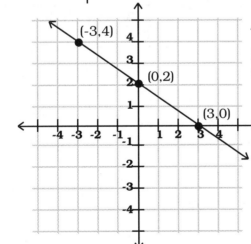

The **y-intercept** is the place where the graph crosses the y-axis. In this example, the y-intercept is (0, 2).
The x-intercept is (3, 0).

Linear Equations, Exercise 5. Graph using zeros.

1. $3x + 5y = 15$
2. $4x - 3y = 12$
3. $x + 2y = -6$
4. $2x - 3y = -4$

9

5. List the x-intercepts in the graphs of problems 1 through 4. List the y-intercepts.

Graphing Horizontal and Verticle Lines.

When an equation contains only the y-variable, its graph is a horizontal line.

When an equation contains only the x-variable, its graph is a vertical line.

•Example 1:

Graph y = 3 .

Solution: This equation means that y is always equal to 3 and the value of x has no affect on y. So if x is 0, this equation says y = 3. If x is -2, the equation still says y = 3.

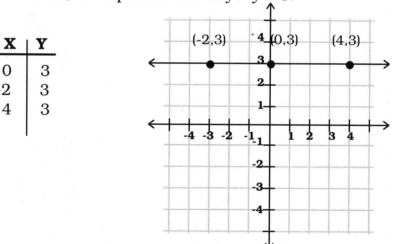

X	Y
0	3
-2	3
4	3

y = 3 is a horizontal line, passing through 3 on the y-axis.

For similar reason, the graph of x = 4 is a vertical line passing through 4 on the x-axis. Another example is also shown below, x = -2.

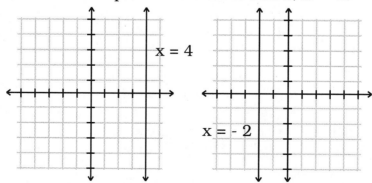

Linear Equations, Exercise 6. Graph the following.

1. y = 2   3. y = -5   5. y - 4 = 0   Hint: add 4 to both sides first.
2. x = 4   4. x = 0   6. x + 1 = 3   Hint: subtract 1 first.

Slope

In the following two sections you will learn the third method of graphing linear equations. This method differs from the previous methods in that the variables are not replaced by numbers. Instead, the third method makes use of a concept called **slope of a line**.

The slope is a number that describes the steepness of the line. The slope is calculated by dividing the rise by the run. The **rise** is the vertical change between any two points on a line. The **run** is the horizontal change between the same two points.

•Example 1:

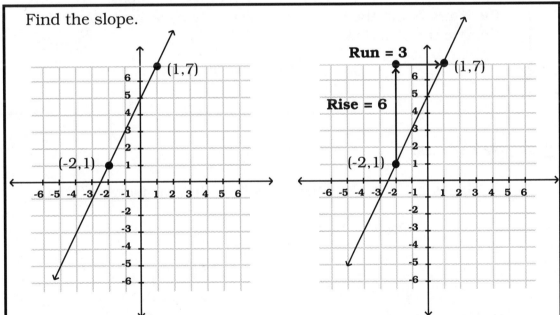

Find the slope.

The rise and the run are found by traveling from one point to the other using only vertical and horizontal pathways. In this example, to travel from point (-2, 1) to the point (1, 7), one would have to move 6 spaces up and then 3 spaces to the right. Therefore, the rise = 6 and the run = 3.

$$\text{Slope} = \frac{\text{rise}}{\text{run}} = \frac{6}{3} = \frac{2}{1} = 2.$$

11

•Example 2:

Find the slope:

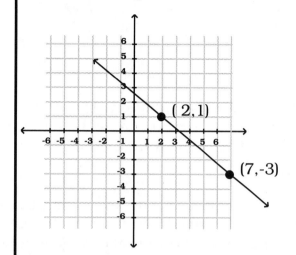 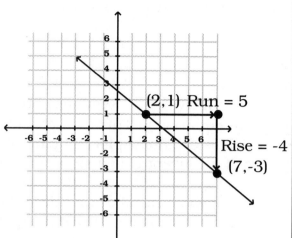

To travel from the point (2, 1) to the point (7, -3) one could move 5 spaces to the right and then down 4 spaces. Therefore, the run = 5 and the rise = -4. The rise is negative to indicate that we traveled downward toward (7, -3):

$$\text{Slope} = \frac{\text{rise}}{\text{run}} = \frac{-4}{+5} = -\frac{4}{5}.$$

Two useful facts:

1. The sign of the slope indicates if the line is rising or falling.

Positive Slope

Negative Slope

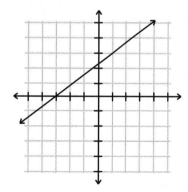 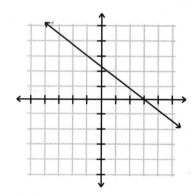

Rises upward from left to right.

Falls downward from left to right.

2. When finding the slope by traveling from one point to another, it makes no difference which point you start with, provided you follow these guidelines:

      a. Traveling vertically upward means the rise is positive;
      b. Traveling vertically downward means the rise is negative;
      c. Traveling to the right means the run is positive;
      d. Traveling to the left means the run is negative.

•Example 3:

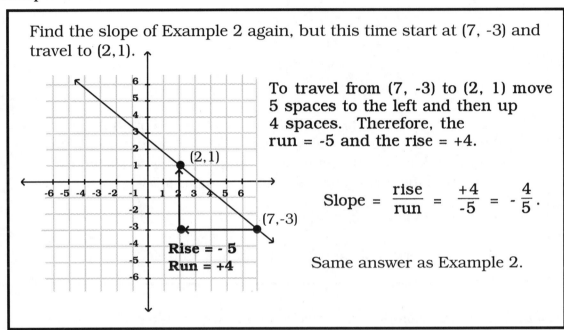

Find the slope of Example 2 again, but this time start at (7, -3) and travel to (2,1).

To travel from (7, -3) to (2, 1) move 5 spaces to the left and then up 4 spaces. Therefore, the run = -5 and the rise = +4.

$$\text{Slope} = \frac{\text{rise}}{\text{run}} = \frac{+4}{-5} = -\frac{4}{5}.$$

Same answer as Example 2.

An alternative method for finding the slope is to use the slope formula.

The Slope Formula: The rise (verticle change) can be found by subtracting the y-coordinates of the two points , and the run (horizontal change) can be found by subtracting the x-coordinates of the two points.

In other words:

$$\text{Slope} = \frac{\text{rise}}{\text{run}} = \frac{y_2 - y_1}{x_2 - x_1}.$$

The following example clarifies the symbols in this formula.

13

•Example 4:

Calculate the slope of Example 2 by using the slope formula.

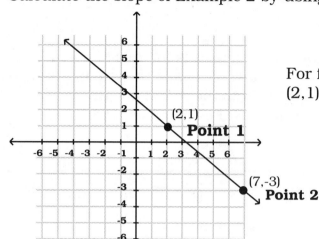

For future reference, we will call (2,1) Point 1 and (7,-3) Point 2.

In the slope formula, the symbol y_2 means the y-coodinate of point two, namely -3. The symbol x_2 means the x-coordinate of Point 2 namely 7. Likewise, the symbols y_1 and x_1, refer to the coordinate of Point 1. Specifically, y_1 means 1 and x_1 means 2.

Therefore: Slope $= \dfrac{y_2 - y_1}{x_2 - x_1}$.

Slope $= \dfrac{-3 - 1}{7 - 2} = \dfrac{-4}{5} = -\dfrac{4}{5}$ as in Example 2.

In Example 4, it makes no difference which point is called Point 1 and which is called Point 2. For instance, if we switch the points and call (2, 1) Point 2 and (7, -3) Point 1, then:

$$\text{Slope} = \frac{1 - (-3)}{2 - 7} = \frac{1 + 3}{-5} = \frac{4}{-5} = -\frac{4}{5}$$

The main idea is to keep the order consistent. That is, the first number in the numerator and the first number in the denominator must come from the same point.

•Example 5:

Find the slope of a line which passes through (2, -3) and (0, 4).

Let (2, -3) be ordered pair 1.
Let (0, 4) be ordered pair 2.

$$\text{Slope} = \frac{y_2 - y_1}{x_2 - x_1} = \frac{4 - (-3)}{0 - 2} = \frac{4 + 3}{-2} = -\frac{7}{2}$$

Linear Equations, Exercise 7. Find the run, rise and slope of each line.

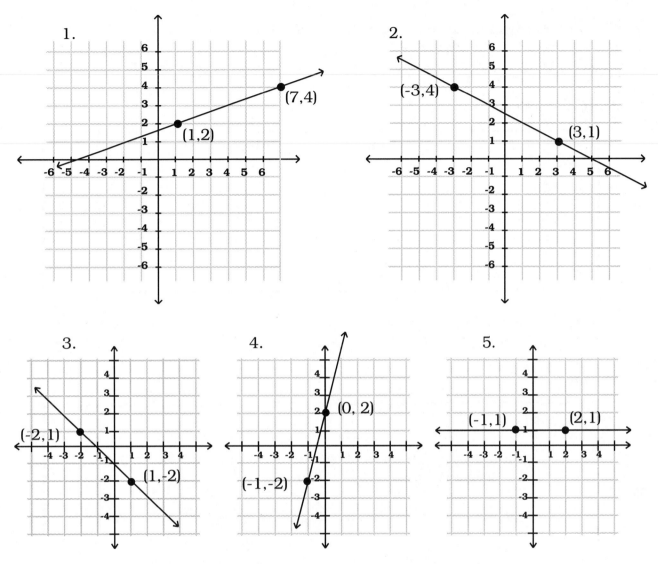

Linear Equations, Exercise 8. Find the slope of a line which:

1. Passes through points (4,5) and (2,3).   Hint: Example 5.
2. Passes through points (0,0) and (3,7).
3. Passes through points (-1,3) and (4,6).
4. Passes through points (0,3) and (3,3).
5. Passes through points (-1,0) and (-5,-6).

Graph Using Slope and Y-Intercept (Method 3)

The y-intercept is the place on the graph where the line crosses
the y-axis (for an example, see Graph Using Zeros, Example 1, page 9.).

In this section, the y-intercept and slope will be used to graph linear

equations. Before learning this method of graphing, you need to understand the slope intercept form of a line.

Slope-Intercept Form: If an equation is of the form $y = mx + b$, then
m = slope and b = y-intercept. For instance, $y = 2x + 3$ is of the form
$y = mx + b$. Therefore its slope equals 2 and its y-intercept equals 3.

•More examples:

	slope	y-intercept
$y = 4x - 7$	4	-7
$y = \frac{1}{3} x + 4$	$\frac{1}{3}$	4
$y = x - 5$	1	-5
$y = -x + 0$	-1	0
$y = mx + b$	m	b

To graph an equation of the form $y = mx + b$:
Plot the y-intercept first. Then use the slope to find other points of the line.

•Example 1:

Graph $y = \frac{3}{2} x - 4$.

Solution: $y = \frac{3}{2} x - 4$ is of the form $y = mx + b$. Therefore,
the slope is $\frac{3}{2}$ and the y-intercept = -4.

Plot the y-intercept by drawing a dot at -4 on the y-axis.

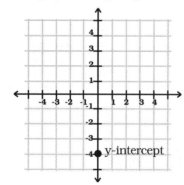

In other words, the y-intercept is one point of the line.

Next the slope is used to find the rise and the run. The rise
equals the numerator of the slope. The run equals the
denominator of the slope.

$$\text{Slope} = \frac{3}{2} = \frac{\text{rise}}{\text{run}} .$$ Therefore rise = 3, run = 2.

Starting at the y-intercept travel up three spaces and then move right 2 spaces to get the second point (Figure A).

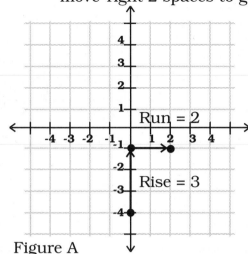

Figure A

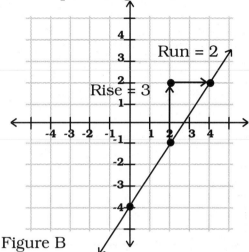

Figure B

To get a third point, rise up 3 spaces from the second point and run over 2. Then draw a line through the three points (Figure B).

•Example 2:

Graph y - 5 = -3x.

Solution: y - 5 = -3x is **not** of the form y = mx + b. Before graphing, we need to get y alone on the left side of the equal sign by adding 5 to both sides:

$$y - 5 = -3x$$
$$y - 5 + 5 = -3x + 5$$
$$y = -3x + 5 \quad \text{which } \textbf{is} \text{ of the form } y = mx + b.$$

Therefore, y-intercept = 5 and slope = -3.

To find the rise and the run, we need to write -3 as a fraction by placing 1 in the denominator:

$$\text{Slope} = -3 = \frac{-3}{1} = \frac{\text{rise}}{\text{run}}. \quad \text{Hence rise} = -3 \text{ and run} = 1.$$

As in the section on slope, a negative rise means we should travel downward. Start at the y-intercept and move down 3 spaces and then right 1 space to get to the next points.

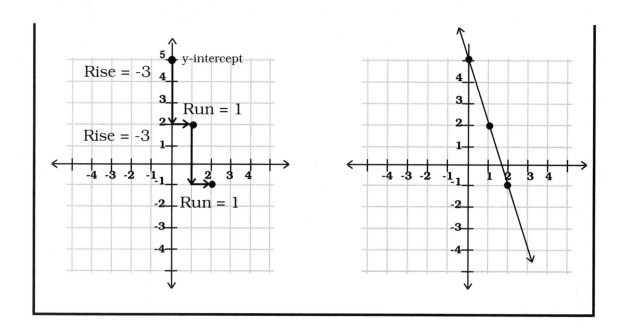

Examples 1 and 2 suggest these steps:

Step 1: Write the equation in the form y = mx + b.
Step 2: Write the slope as a fraction: rise = numerator,
run = denominator.
Step 3: Plot the y-intercept first, then use the rise and the run to get
the next point.

•Example 3:

Graph 4x - 5y = 10 .

Step 1: Get y by itself on one side of the equal sign.

4x - 5y = 10

$\quad$ -5y = -4x + 10 $\qquad$ Subtract 4x both sides.

$\quad$ (-5y) = (-4x + 10) $\qquad$ Enclose in parentheses.

$\frac{1}{5}$(-5y) = $\frac{1}{5}$(-4x + 10) $\qquad$ Multiply by reciprocal of -5.

$\frac{1}{5}$(-5y) = - $\frac{1}{5}$(4x) - $\frac{1}{5}$. 10 $\qquad$ Distribute - $\frac{1}{5}$

$\frac{1}{5}$ ($\frac{-5}{1}$ y) = - $\frac{5}{1}$ ($\frac{-4}{1}$ x) - $\frac{1}{5}$. $\frac{10}{1}$ $\qquad$ Since -5 = - $\frac{5}{1}$, etc.

$\quad\quad$ y = $\frac{4}{5}$x - 2 $\qquad$ Cancel the 5s and negatives.

Hence y-intercept = -2, slope = $\frac{4}{5}$, rise = 4, run = 5.

18

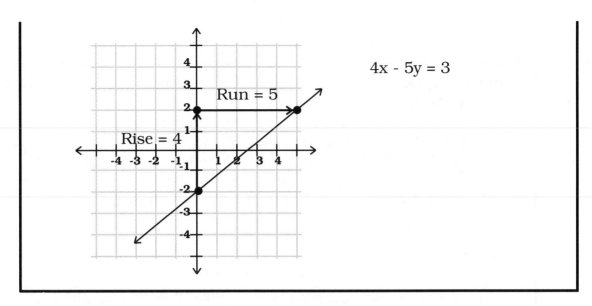

$4x - 5y = 3$

Linear Equations, Exercise 9. Indentify the y-intercept and slope, then graph.

1. $y = \frac{3}{2}x - 1$

2. $y = -\frac{1}{3}x + 2$ Hint: $-\frac{1}{3} = \frac{-1}{+3} = \frac{\text{rise}}{\text{run}}$.

3. $y = 2x - 4$ Hint: $2 = \frac{2}{1} = \frac{\text{rise}}{\text{run}}$.

4. $y = -2x + 4$

5. $y = \frac{1}{2}x$ Hint: This is the same as $y = \frac{1}{2}x + 0$, so y-intercept = 0.

6. $y = x + 2$ Hint: This is the same as $y = 1x + 2$ and $1 = \frac{1}{1} = \frac{\text{rise}}{\text{run}}$.

7. $y = -x + 2$ Hint: -x means -2x and $-1 = \frac{-1}{1} = \frac{\text{rise}}{\text{run}}$.

8. $y = \frac{4}{5}x - 3$

For 9 - 12, solve for y, then graph. For 13 - 16, solve for y to determine intercept and slope, but don't graph.:

9. $y - 2 = \frac{1}{3}x$ 13. $2x + 5y = 15$ Hint: Follow Example 3.

10. $y - 2x = 1$ 14. $-y = 2x - 5$ Hint: Multiply both sides by -1.

11. $\frac{1}{3} + y = 0$ 15. $-2x - y = 5$

12. $4x + 3y = 0$ 16. $x - 3y = -2$

Writing Equations

If the equation of a line is not known, it can sometimes be found. More specifically, we can find the equation provided we know two pieces of information:

     1. The slope of the line;
     2. Any one point on the line.

•Example 1:

> Find the equation of the line that passes through the point (-2,1) and has a slope of 3.
>
> Solution: to find the equation, we will use the point-slope form of a line, $y = mx + b$. Our goal is to find the m and the b. Remember from the previous section that m = slope. We are given the slope is 3. Therefore, we can replace m by 3:
>
> $$y = mx + b$$
> $$y = 3x + b$$
>
> To find b, use the given point . We are given that the line passes through (-2,1). Therefore, (-2,1) is a solution to the equation which means we can replace x by -2 and y by 1 to get:
>
> $$y = 3x + b$$
> $$1 = 3(-2) + b$$
> $$1 = -6 + b$$
> $$7 = b$$
>
> Therefore, the equation of the line is $y = 3x + 7$, because $m = 3$ and $b = 7$.

Example 1 suggests these steps:

Step 1: In $y = mx + b$, replace m by the given slope.
Step 2: Replace x and y with the ordered pair and solve for b.

Linear Equations, Exercise 10. Find the equation of a line which:

1. Passes through (3,4) and has a slope of 1;
2. Passes through (6,-5) and has a slope of 2;
3. Passes through (-2,1) and has a slope of -4;
4. Passes through (6,0) and has a slope of -1;
5. Passes through (0,0) and has a slope of 1.
6. Given a line that passes through (1,2) and (2,10):
   a. Find the slope using Slope = $\dfrac{y_2 - y_1}{x_2 - x_1}$
   b. In $y = mx + b$ replace m by the answer to Part a. Also replace x and y by the ordered pair (1,2).
   c. Solve for b.

Systems of Equations

In this chapter you will learn three methods for solving systems of linear equations.

Systems and Solutions

A **system of equations** is a collection of two or more equations.
For example:

$$x + 3y = 4$$
$$2x + y = -7 \qquad \text{is a system of two equations.}$$

A **solution** to a system of two equations is any ordered pair that makes <u>both</u> equations true. For example (-5,3) is a solution to the system shown above. Replacing x with -5 and y with 3 in <u>both</u> equations gives:

$x + 3y = 4$		$2x + y = -7$	
$-5 + 3(3) = 4$		$2(-5) + 3 = -7$	
$-5 + 9 = 4$		$-10 + 3 = -7$	
$4 = 4$	True.	$-7 = -7$	True.

Hence (-5,3) is a solution to the system.

Any ordered pair that does not lead to a true statement in both equations is **not** a solution. For instance, (4,0) is not a solution to the above system because replacing x with 4 and y with 0 gives:

$x + 3y = 4$		$2x + 4 = -7$	
$4 + 3(0) = 4$		$2(4) + 0 = -7$	
$4 + 0 = 4$		$8 + 0 = -7$	
$4 = 4$	True.	$8 = -7$	False.

(4,0) leads to a false statement in the second equation. Therefore, (4,0) is **not** a solution to the system.

Systems of Equations, Exercise 1. Tell whether or not each ordered pair is a solution to the given system.

1. $x + y = 4$
   $2x + y = 5$
   a. $(1,3)$ b. $(0,4)$ c. $(2,1)$

2. $5x + y = 11$
   $4x - 6 = 2y$
   a. $(1,6)$ b. $(2,1)$ c. $(0,11)$

3. $-2x + y = 6$
   $x - y = -3$
   a. $(0,3)$ b. $(-2,2)$ c. $(-3,0)$

4. $x = 5y - 2$
   $10y = 2x + 4$
   a. $(3,1)$ b. $(-2,0)$ c. $(8,2)$

5. $y = 7x + 9$
   $y = 3x + 1$
   a. $(0,9)$ b. $(-2,-5)$ c. $(1,1)$

Solving Systems by Graphing (Method 1)

The solution(s) to a system of equations can be found by graphing each equation on the same x, y axes. Any point which is common to both lines is a solution to the system.

•Example 1:

> Find the solution to the system:
>
> $2x + y = 4$
> $x - 3y = -5$
>
> Procedure: Start with the first equation, $2x + y = 4$. Graph this equation by the point plotting method of Chapter 1.

When x = 0

$2x + y = 4$

$2(0) + y = 4$

$0 + y = 4$

$y = 4$

X	Y
0	4
2	0
3	-2

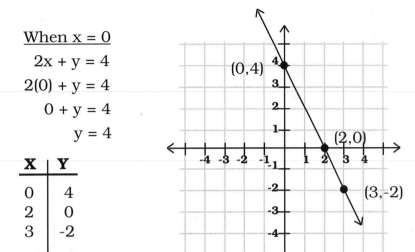

On the same axes, use the point plotting method to graph the second equation, x - 3y = -5.

When x = -2

$x - 3y = -5$

$-2 - 3y = -5$

$-3y = -5 + 2$

$-3y = -3$

$y = 1$

X	Y
-2	1
4	3
-5	0

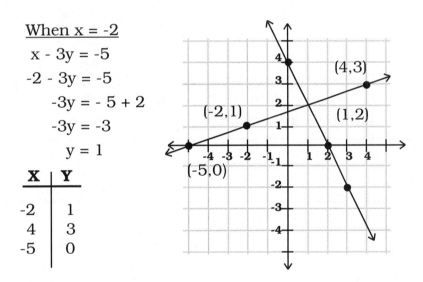

These lines cross each other at (1,2). Therefore (1,2) is the solution to the system. If the lines cross, we say the system is **consistent**.

The solution in the above example can be checked (see Exercise 15).

•Example 2:

Find the solution to the system:

$$y = 2x + 3$$
$$8x - 4y = 4$$

Solution: Graph both lines on the same axes using the point plotting method.

$y = 2x + 3$

X	Y
0	3
1	5
-2	-1

$8x - 4y = 4$

X	Y
1	1
2	3
0	-1

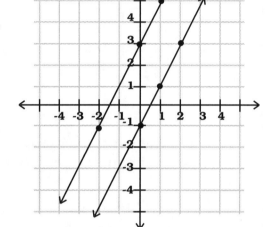

These lines never cross each other. Lines that do not cross are called parallel lines. There is no solution, and we say the system is **inconsistent**.

•Example 3:

Find the solution to the system:

$$x + 2y = 2$$
$$3x + 6y = 6$$

$x + 2y = 2$

X	Y
0	1
2	0
4	-1

$3x + 6y = 6$

X	Y
0	1
2	0
4	-1

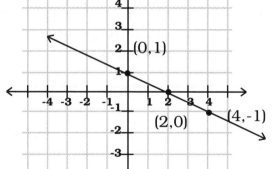

Both lines occupy the same space.

When graphed, these equations turn out to be exactly the same line. Hence, any point on the line is a solution. There are an infinite number of points on the line. Therefore, there are an infinite number of solutions. When the lines are identical, we say the system is **dependent**.

Summary of Solving by Graphing

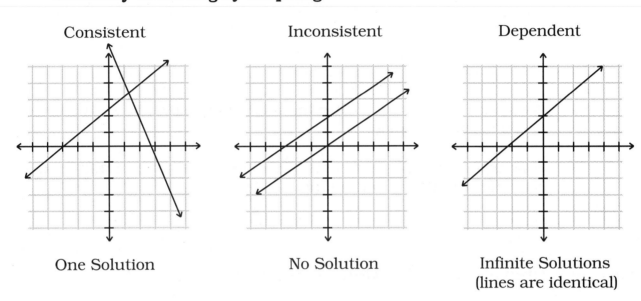

Consistent

One Solution

Inconsistent

No Solution

Dependent

Infinite Solutions
(lines are identical)

Systems of Equations, Exercise 2. Label each as consistent, inconsistent, or dependent.

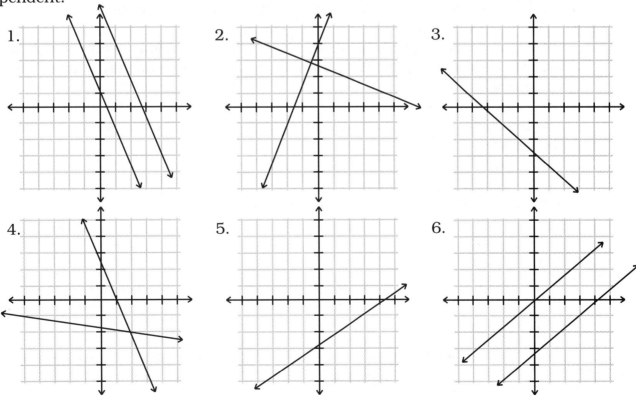

1.

2.

3.

4.

5.

6.

Graph both lines on the same x,y-axes. State whether the system is consistent, inconsistent, or dependent. If the system is consistent, find the solution.

7. $x + y = 1$
    $x - y = -3$

8. $x + y = 1$
    $x + y = 2$

9. $x + y = 1$
    $2x + 2y = 2$

10. $y = 3x + 1$
    $y = 2x + 2$

11. $y = 2x + 1$
    $2y = 4x + 4$

12. $y = 5x$
    $4y = x$

13. $2y = x - 1$
    $2x = 4y + 2$

14. $y = 2 + x$
    $x + 2y = 1$

15. In Example 1, verify that (1,2) is the solution by replacing x with 1 and y with 2 in each equation.

16. In Example 3, multiply both sides of the first equation by 3 and compare this result with the second equation.

Solving Systems by Substitution (Method 2)

The graphing method enables us to visualize the solutions to a system. However, graphing is time consuming and sometimes inaccurate. In this section you will learn an alternative method called **substitution**.

We will use exactly the same examples as in the previous section to show how substitution can indicate whether a system is consistent, inconsistent, or dependent.

The first example of the graphing section showed that (1,2) is the solution
to: $2x + y = 4$
    $x - 3y = -5$.

We can also find this solution by the substitution method.

•Example 1:

Solve $2x + y = 4$
       $x - 3y = -5$.

Solution: In either equation, get x or y alone (isolated) on one side of the equal sign. Let's choose to isolate x of the second

26

equation by adding 3y to both sides:

$$x - 3y = -5$$
$$x - 3y + 3y = -5 + 3y$$
$$x = -5 + 3y \qquad \text{Since } -3y + 3y = 0.$$

$x = -5 + 3y$ says that x is equal to -5 + 3y. Since the expression -5 + 3y is equal to x, it can be used as a substitute (or replacement) for x.

In the other equation, 2x + y = 4, replace x with -5 + 3y to get:

$$2x + y = 4$$
$$2(-5 + 3y) + y = 4 \qquad \text{Substituting } -5 + 3y \text{ for } x.$$
$$-10 + 6y + y = 4 \qquad \text{Distribute the 2.}$$
$$-10 + 7y = 4 \qquad \text{Collect terms.}$$
$$7y = 14 \qquad \text{Add 10 to both sides.}$$
$$y = 2$$

$y = 2$ means that 2 is the y-coordinate of the solution. To find the x-coordinate of the solution, replace y with 2 in either original equation. Let's use the original equation x - 3y = -5:

$$x - 3y = -5$$
$$x - 3(2) = -5 \qquad \text{Replace } y \text{ with 2.}$$
$$x - 6 = -5$$
$$x = 1 \qquad \text{By adding 6 to both sides.}$$

Hence, the solution is (1,2). Note this is the same solution as found by graphing.

The above example suggest these steps:

To Solve by Substitution:

Step 1: In either equation, isolate x or y.
Step 2: Into the other equation, substitute the result of Step 1.
Solve for one coordinate of the solution.
Step 3: Into either original equation, replace the appropriate letter by the coordinate in Step 2.

In the second example of the graphing method, we saw that the system $\quad y = 2x + 3 \quad$ was inconsistent.
$$8x - 4y = 4$$

Let's see what happens when we try to solve this system by the substitution method.

•Example 2:

Solve $\quad y = 2x + 3$
$\qquad 8x - 4y = 4.$

Solution: Step 1 is already done, since y is isolated in the top equation.

Step 2:	$8x - 4y = 4$	
	$8x - 4(2x + 3) = 4$	Substitute $2x + 3$ in for y.
	$8x - 8x - 12 = 4$	Distribute the -4.
	$0 - 12 = 4$	Since $8x - 8x = 0$
	$-12 = 4$	False.

A false statement means that there is no solution, the system is inconsistent, and the lines are parallel.

Three Useful Facts

1. The system is <u>consistent</u> if substitution results in a specific solution like (1,2), x = 1, y = 2.
2. The system is <u>inconsistent</u> if substitution results in a false statement like -12 = 4.
3. The system is <u>dependent</u> if substitution results in a true statement without variables like 0 = 0 or 5 = 5.

•Example 3

Solve $\quad x + 2y = 2$
$\qquad 3x + 6y = 6.$

Solution: The most convenient variable to isolate is the x in the first equation, since all the other variables are attached to numbers.

Step 1: $x + 2y = 2$

          $x = 2 - 2y$ Subtract 2y from both sides.

Step 2: $3x + 6y = 6$ (second equation)

          $3(2 - 2y) + 6y = 6$ Substitute 2 - 2y for x.

          $6 - 6y + 6y = 6$ Distribute the 3.

          $6 - 0 = 6$ Since -6y + 6y = 0.

          $6 = 6$ True.

Therefore, the system is dependent. Note, this system is the same as the system of Example 3 in the graphing section.

•Example 4:

Solve $2x - 3y = 9$
      $6x + 2y = 5$.

Step 1: Choose a variable to isolate. There is no obvious choice. Let's try to isolate the x in the first equation.

$$2x - 3y = 9$$
$$2x = 9 + 3y \quad \text{Add 3y to both sides.}$$
$$(2x) = (9 + 3y) \quad \text{Enclose in parentheses.}$$
$$\tfrac{1}{2}(2x) = \tfrac{1}{2}(9 + 3y) \quad \text{Multiply by } \tfrac{1}{2}$$
$$\tfrac{1}{2} \cdot \tfrac{2}{1}x = \tfrac{1}{2} \cdot \tfrac{9}{1} + \tfrac{1}{2} \cdot \tfrac{3}{1}y \quad \text{Distribute } \tfrac{1}{2}$$
$$x = \tfrac{9}{2} + \tfrac{3}{2}y$$

Step 2: $6x + 2y = 5$

$$6(\tfrac{9}{2} + \tfrac{3}{2}y) + 2y = 5 \quad \text{Substitute for x.}$$
$$\tfrac{6}{1} \cdot \tfrac{9}{2} + \tfrac{6}{1} \cdot \tfrac{3}{2}y + 2y = 5 \quad \text{Distribute } \tfrac{6}{1}$$
$$27 + 9y + 2y = 5 \quad \text{Cancel denominators.}$$
$$11y = -22 \quad \text{Subtract 27 from both sides.}$$
$$y = -2$$

Step 3: $2x - 3y = 9$

          $2x - 3(-2) = 9$ Replace y with -2.

          $2x + 6 = 9$

          $x = \tfrac{3}{2}$ Subtract 6, divide by 2.

Solution $(\frac{3}{2}, -2)$.	A consistent system.

The next section offers a better way to do Example 4.

Systems of Equations, Exercise 3. Solve by substitution.

In problems 1, 2, and 3, start by isolating the y in the second equation.

1.	$3x + 2y = -1$	2.	$2x + 3y = 1$	3.	$3x + 2y = 4$
	$-4x + y = 5$		$2x + y = 3$		$-4x + y = 2$

In each system, isolate **any** variable, then solve by substitution..

4.	$x = y - 7$	5.	$x - y = 5$	6.	$x = 1 - y$
	$x + 2y = 5$		$2x + y = 1$		$x + y = 2$
	Hint: x already isolated.				
7.	$2x + y = 4$	8.	$x - 4y = 7$	9.	$x + y = 1$
	$3x - y = 1$		$2x + y = 5$		$2x + 2y = 2$
10.	$8x + y = 1$	11.	$x = y + 3$	12.	$x - 7y = 0$
	$9x + 2y = 2$		$7x - 7y = 3$		$x + y = -8$
13.	$y = 3x + 2$	14.	$5x + 9y = 0$	15.	$2x - 5y = 4$
	$4x + 5 = y$		$4x - 3y = 0$		$4x + 3y = -5$
			Hint: See Example 4.		

16. If substitution results in a false statement, then the system
    is _____.

17. If substitution results in a statement like 4 = 4, then the system
    is _____.

18. In Example 4, check the solution in both equations.

Solve Using Addition (Method 3)

In this section you will learn the third method of solving systems of equations called the **addition method**.

This method involves adding the equations together in order to eliminate x or y.

•Example 1:

Solve $3x + y = 1$
$x - y = 7$.

Solution: When these equations are added together, the y-variable cancels out.

$$3x + y = 1$$
$$+ \quad \underline{x - y = 7}$$
$$4x + 0 = 8$$
$$4x = 8$$
$$x = 2$$

Since $y + (-y) = 0y = 0$,
$3x + x = 4x$, and $1 + 7 = 8$.

$x = 2$ means the solution has x-coordinate of 2. To find the y-coordinate, replace x with 2 in either original equation.

$$3x + y = 1$$
$$3(2) + y = 1$$
$$6 + y = 1$$
$$y = -5$$

The solution is $(2, -5)$.

The solution in the above example can be checked (see Exercise 12).

•Example 2:

Solve $2x + y = 4$
$x - 3y = -5$.

Solution: Our goal is to eliminate x or y. Neither x nor y would be eliminated if we were to add the equations as they now appear. We can remedy this situation, however, by multiplying both sides of the first equation by 3 to get:

$$\begin{array}{ccc} 2x + y = 4 \\ x - 3y = -5 \end{array} \longrightarrow \begin{array}{c} 3(2x + y) = 3 \cdot 4 \\ x - 3y = -5 \end{array} \longrightarrow \begin{array}{c} 6x + 3y = 12 \\ x - 3y = -5 \end{array}$$

Now, when these equations are added together, the y-terms cancel out:

$$6x + 3y = 12$$
$$+ \quad \underline{x - 3y = -5}$$
$$7x + 0 = 7$$
$$7x = 7$$
$$x = 1$$

Since $6x + x = 7x$, $3y + (-3y) = 0$,
and $12 + (-5) = 7$.

31

Replace x with 1 in either original equation to get:

$$2x + y = 4$$
$$2(1) + y = 4$$
$$y = 2$$

Hence, the solution is (1,2).

Notice that this is the same example and solution as Example 1 of the *Solve by Graphing* and Example 1 of *Solve by Substitution* sections.

In the above example, we decided to multiply the first equation by +3, because the y-term in the second equation had a coefficient of -3. In other words, we needed to make the coefficients opposite so they would cancel out.

To Solve by the Addition Method
Step 1: If necessary, rearrange the equation to get x and y on the same side of the equal sign.
Step 2: If necessary, multiply one (or both) equations to make the coefficient of the x or y terms opposite.
Step 3: Add equations. Solve for a coordinate of the solution.
Step 4: In either original equation, substitute the results of step 3.

•Example 3:

Solve $5y = 2 - x$
   $x + 4y = 1.$

Step 1: In the first equation, move the x term to the left of the equal sign by adding x to both sides:

$$5y = 2 - x \longrightarrow x + 5y = 2$$
$$x + 4y = 1 \qquad\qquad x + 4y = 1$$

Step 2: The x term can be made opposite by multiplying both sides of either equation by -1. Let's multiply the second equation by -1.

$$x + 5y = 2 \longrightarrow x + 5y = 2$$
$$-1 (x + 4y) = -1 \cdot 1 \qquad -x - 4y = -1$$

32

Step 3:   x + 5y = 2   Step 4:   x + 4y = 1
        + -x - 4y = -1   x + 4(1) = 1
        ─────────────   x + 4 = 1
                y = 1   x = -3

Solution = (-3,1)

•Example 4

Solve 2x + 3y = 1
      5x + 2y = -3.

Step 1: Variables are already on the left side of the equal sign.
Step 2: Let's choose to make the coefficients of the x terms
opposite. Multiply the first equation by -5. Multiply the
second equation by +2 to get:

2x + 3y = 1 ⟶ -5(2x + 3y) = -5(1) ⟶ -10x - 15y = -5
5x + 2y = -3   2(5x + 2y) = 2(-3)   10x + 4y = -6

Step 3:   -10x - 15y = -5   Step 4:   2x + 3y = 1
        +   10x + 4y = -6   2x + 3(1) = 1
        ──────────────   2x + 3 = 1
              -11y = -11   2x = -2
                 y = 1   x = -1

Solution = (-1,1)

•Example 5:

Use addition to show that   y = 2x + 3   is inconsistent.
                             8x - 4y = 4

Step 1:   y = 2x + 3   ⟶   -2x + y = 3   (Subtract 2x)
           8x - 4y = 4   8x - 4y = 4

Step 2:   4(-2x + y) = 4 · 3 ⟶   -8x + 4y = 12
               8x - 4y = 4   8x - 4y = 4

Step 3:   -8x + 4y = 12
        +   8x - 4y = 4
        ──────────────
               0 = 16   False. Hence, inconsistent.

Notice the above example is identical to *Solve Systems by Graphing*, Example 2 and *Solve Systems by Substitution*, Example 2.

The chart below summarizes the results of the three methods when the system is consistent, inconsistent, or dependent.

In either Addition or Substitution Method:	**Consistent** x = number y = number	**Inconsistent** False Statement, like 0 = 12.	**Dependent** True Statement, like 0 = 0.

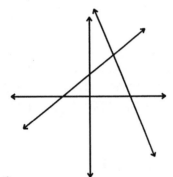 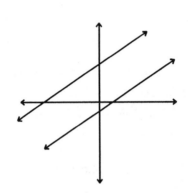 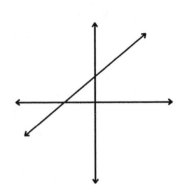

Graphing Method:	**Consistent**	**Inconsistent**	**Dependent**

Systems of Equations, Exercise 4.   Solve by the addition method.

1. x + y = 1
    x - y = -3

2. -x + 2y = 1
    x + 7y = -10

3. x - 4y = 13
    2x + 4y = 2

4. x - 2y = 0
    -x + 2y = 4

5. -x + y = 0
    x = 3 - 2y
    Hint: Example 3, Step 1

6. x +3y = 5
    x + y = 3
    Hint: Multiply either equation by -1.

7. x - y = 4
    3x + 2y = 2
    Hint: Multiply first equation by 2.

8. x - y = 3
    -2x + 2y = -6

9. 2x + y = 5
    5x + 3y = 7
    Hint: Multiply first equation by -3.

10. 2x + 2y = 0
     5x + 3y = 2
     Hint: Example 4.

11. 3x = 1 - 4y
     2x + 5y = 3

12. Check the solution in Example 1 by replacing x by 2 and y by -5 in both equations.

13. Do Example 4 in *Solve By Substitution* but use the addition method.

34

Rational Expressions

In this chapter we will assume that the student is already familiar with factoring polynomials. For an explanation of how to factor polynomials see Algebra, Book 1, or any similar book.

This chapter is not essential for understanding the next two.

Rational Expressions

A **rational expression** is any fraction with a polynomial in its numerator and a polynomial in its denominator. For instance:

$$\frac{15}{35} \qquad \frac{x}{xy} \qquad \frac{2}{2 + 6x} \qquad \frac{x^2 + 3x + 2}{xy + 2y}$$

Rational expressions behave like fractions in that they can be reduced, added together, subtracted, multiplied, or divided. Before learning these procedures, however, you need to understand the difference between a term and a factor.

Terms and Factors

Numbers or variables connected to each other by addition or subtraction are called **terms**. For instance, in 5 + 2, the terms are 5 and 2.

Numbers or variables connected to each other by multiplication are called **factors**. For instance, in 5 · 7, the factors are 5 and 7.

•More Examples:

> In 2 + x, the terms are 2 and x.
> In x^2 - 3x + 2, the terms are x^2 , -3x and 2.
>
> In 2x, the factors are 2 and x.
> In 3(x + 1), the factors are 3 and (x + 1).
> In (y + 2) (y - 3), the factors are (y + 2) and (y - 3).

Having made the distinction between terms and factors, we need to state the cancellation rule for fractions.

Cancellation Rule: It is legal to cancel any <u>factor</u> that appears in both the numerator and denominator. However, <u>terms cannot</u> be cancelled.

For instance, in the fraction $\frac{5 \cdot 7}{5}$, it is legal to cancel the 5s to get $\frac{5 \cdot 7}{5} = 7$, because 5 is a factor in the numerator and the denominator.

However, in the fraction $\frac{5+2}{5}$, it is wrong to cancel the 5s because 5 is a term in the numerator. In other words, $\frac{5+2}{5} = \frac{7}{5}$. Cancelling the 5s would result in $\frac{\cancel{5}+2}{\cancel{5}} = 2$, which is not $\frac{7}{5}$.

The above cancelling rule makes it possible to reduce rational expressions.

To reduce Rational Expressions

Step 1: Factor the numerator and denominator.
Step 2: Cancel factors that appear in both the numerator and denominator.

•Example 1:

Reduce $\frac{30}{42}$.

Step 1: $\frac{30}{42} = \frac{2 \cdot 3 \cdot 5}{2 \cdot 3 \cdot 7}$

Step 2: $\qquad = \frac{\cancel{2} \cdot \cancel{3} \cdot 5}{\cancel{2} \cdot \cancel{3} \cdot 7} = \frac{5}{7}$

•Example 2:

Reduce $\frac{2}{2 + 6x}$.

Step 1: $\frac{2}{2 + 6x} = \frac{2}{2(1 + 3x)}$

Step 2: $\qquad = \frac{\overset{1}{\cancel{2}}}{\underset{1}{\cancel{2}}(1 + 3x)} = \frac{1}{1 + 3x}$

In the above example, it is legal to cancel the 2s after the denominator has been factored, since 2 and (1 + 3x) are factors.

It would be illegal, however, to cancel out the 2s before the denominator was factored:

Incorrect Way

$$\frac{\cancel{2}}{2 + 6x} = \frac{\cancel{2}}{\cancel{2} + 6x}$$
$$= \frac{1}{6x}$$

Correct Way

$$\frac{2}{2 + 6x} = \frac{2}{2(1 + 3x)} \quad \text{Factor first.}$$
$$= \frac{1}{1 + 3x} \quad \text{Then cancel.}$$

The *incorrect way* is wrong because the 2 in the denominator is a term, not a factor.

•Example 3:

> Reduce $\dfrac{x^2 - 2x}{x^2 - 3x + 2}$.
>
> Step 1: $\quad \dfrac{x^2 - 2x}{x^2 - 3x + 2} = \dfrac{x(x - 2)}{(x - 2)(x - 1)}$
>
> Step 2: $\quad\quad\quad = \dfrac{x\cancel{(x - 2)}}{\cancel{(x - 2)}(x - 1)} = \dfrac{x}{x - 1}$

In the above example, it is legal to cancel out the (x - 2) factors.

It would be illegal, however, to begin Example 3 by cancelling out each x^2.

Incorrect Way

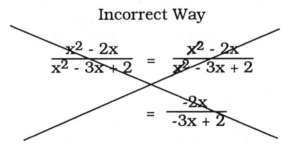

The incorrect way is wrong, since the x^2 is a term, not a factor.

Rational Expressions, Exercise 1. In each sentence fill in the blank with the word *terms* or *factors*.

1. In 2 + 3x, 2 and 3x are_____.
2. In $y^2 - 2y + 1$, y^2, -2y and +1 are _____.
3. In 2(3x), 2 and 3x are _____.
4. In 7(x - 2), 7 and (x - 2) are _____.
5. In x(x - 3), x and (x - 3) are _____.
6. In (y + 1)(y - 2), (y + 1) and (y - 2) are _____.
7. In a fraction, it is legal to cancel_____ , but illegal to cancel _____.

Reduce by cancelling common factors. Some will not reduce.

8. $\dfrac{7 \cdot 3}{7}$ 9. $\dfrac{x \cdot y}{x \cdot z}$ 10. $\dfrac{3(y + 2)}{4(y + 2)}$ 11. $\dfrac{x(x + 1)}{5x}$

12. $\dfrac{y + 3}{3}$ 13. $\dfrac{(2x + 1)}{x(2x + 1)}$ 14. $\dfrac{x(y - 2)}{(y - 2)(y + 7)}$ 15. $\dfrac{2x + 1}{2x}$

Reduce. Remember to factor numerator and denominator before cancelling.
Hint: Some will not reduce.

16. $\dfrac{15}{20}$ 17. $\dfrac{3}{9}$ 18. $\dfrac{6}{35}$ 19. $\dfrac{2x + 4}{2}$ 20. $\dfrac{3y^2 + 12}{3y + 6}$

21. $\dfrac{3x + 3}{4}$ 22. $\dfrac{x + 3}{x^2 + 3x}$ 23. $\dfrac{xy + 2y}{xy + 3y}$

24. $\dfrac{4x + 8}{x}$ 25. $\dfrac{x^2 + 3x + 2}{x + 1}$ 26. $\dfrac{x^2 + 5x + 6}{x^2 + 4x + 4}$

27. $\dfrac{y^2 - 9}{y^2 - 6y + 9}$ 28. $\dfrac{x^2 + 5x + 6}{x^2 + 2x + 1}$ 29. $\dfrac{x^2 + 2xy + y^2}{x^2 - y^2}$

Multiplying and Dividing Rational Expressions

Often in math or science a situation arises that requires us to multiply or divide two rational expressions together. The process of multiplying or dividing is similar to the process of reducing in the previous section.

To Multiply:

Step 1: Factor the numerators and denominators.
Step 2: Cancel common factors.
Step 3: Multiply numerators together. Multiply denominators.

•Example 1:

Multiply $\dfrac{14}{3} \cdot \dfrac{5}{21}$.

Step 1: $\dfrac{14}{3} \cdot \dfrac{5}{21} = \dfrac{2 \cdot 7}{3} \dfrac{5}{7 \cdot 3}$

Step 2: $= \dfrac{2 \cdot \cancel{7}}{3} \dfrac{5}{\cancel{7} \cdot 3}$

Step 3: $= \dfrac{10}{9}$ Since $2 \cdot 5 = 10,\ 3 \cdot 3 = 9.$

38

•Example 2:

Multiply $\dfrac{2x + 10}{x} \cdot \dfrac{6}{3x + 15}$.

$= \dfrac{2(x + 5)}{x} \cdot \dfrac{3 \cdot 2}{3(x + 5)}$ Step 1.

$= \dfrac{2(\cancel{x + 5})}{x} \cdot \dfrac{\cancel{3} \cdot 2}{\cancel{3}(\cancel{x + 5})}$ Step 2.

$= \dfrac{4}{x}$

To Divide:

Step 1: Flip the second fraction.
Step 2: Multiply as above.

•Example 3:

Divide $\dfrac{21}{6} \div \dfrac{14}{5}$

$= \dfrac{21}{6} \cdot \dfrac{5}{14}$ Flip second fraction.

$= \dfrac{3 \cdot 7}{2 \cdot 3} \cdot \dfrac{5}{2 \cdot 7} = \dfrac{\cancel{3} \cdot \cancel{7}}{2 \cdot \cancel{3}} \cdot \dfrac{5}{2 \cdot \cancel{7}} = \dfrac{5}{4}$

•Example 4:

Divide $\dfrac{y - 8}{y^2 - 4} \div \dfrac{3y}{y^2 + 2y}$.

$= \dfrac{y - 8}{y^2 - 4} \cdot \dfrac{y^2 + 2y}{3y}$ Flip second fraction.

$= \dfrac{y - 8}{(y + 2)(y - 2)} \cdot \dfrac{y(y + 2)}{3y} = \dfrac{y - 8}{(\cancel{y + 2})(y - 2)} \cdot \dfrac{\cancel{y}(\cancel{y + 2})}{3\cancel{y}}$

$= \dfrac{y - 8}{3(y - 2)} = \dfrac{y - 8}{3y - 6}$

Rational Expressions, Exercise 2.

Multiply	Divide

1. $\dfrac{14}{15} \cdot \dfrac{35}{8}$

2. $\dfrac{x \cdot x}{y} \cdot \dfrac{y}{x}$

3. $\dfrac{x(x + 2)}{y} \cdot \dfrac{y \cdot y}{x + 2}$

4. $\dfrac{x^2 + 3x}{x + 5} \cdot \dfrac{2x + 10}{x}$

5. $\dfrac{x^2 + x - 6}{x} \cdot \dfrac{x^2 - 3x}{x^2 - 9}$

6. $\dfrac{10}{27} \div \dfrac{25}{9}$

7. $\dfrac{x}{y} \div \dfrac{x}{y}$

8. $\dfrac{x^2 - 5x - 6}{x - 6} \div \dfrac{x^2 + 2x + 1}{x}$

9. $\dfrac{x^2 - 6xy + 8y^2}{x^2 - 7xy + 12y^2} \div \dfrac{1}{(x - 3y)}$

Least Common Multiple

In the next section, you will learn how to add or subtract two rational expressions. As in adding or subtracting two fractions, this will involve getting a common denominator. A concept that will enable you to find a common denominator is called the **least common multiple**.

The **least common multiple** (LCM) of two numbers is the smallest number that both numbers can divide into evenly.

For instance, the LCM of 6 and 9 is 18, since 18 is the smallest number that both 6 and 9 can divide into evenly.

There is a method for finding the LCM.

To Find the LCM of Two Numbers
Step 1: Factor each number into primes.
Step 2: For each prime factor, note the greatest number of times it appears in either number.
Step 3: Multiply the primes of Step 2 together.

•Example 1:

Find the LCM of 30 and 24.

Step 1:

$$30$$
$$= 5 \cdot 6$$
$$= 5 \cdot 2 \cdot 3$$

$$24$$
$$= 4 \cdot 6$$
$$= 2 \cdot 2 \cdot 2 \cdot 3$$

That is, $30 = 5 \cdot 2 \cdot 3$ and $24 = 2 \cdot 2 \cdot 2 \cdot 3$.

Step 2: To find the LCM, ask yourself these questions:
   a. Which expression in Step 1 contains the most 2s?
      Answer: $24 = 2 \cdot 2 \cdot 2 \cdot 3$ contains more 2s than $30 = 5 \cdot 2 \cdot 3$.
      Specifically, $24 = 2 \cdot 2 \cdot 2 \cdot 3$ contains three 2s. This
      means that the LCM will contain three 2s.

   b. Which expression contains the most 5s?
      Answer: The factor 5 appears most often in $30 = 5 \cdot 2 \cdot 3$.
      Since $30 = 5 \cdot 2 \cdot 3$ contains one 5, so will the LCM.

   c. Which expression contains the most 3s?
      Answer: Both $30 = 5 \cdot 2 \cdot 3$ and $24 = 2 \cdot 2 \cdot 2 \cdot 3$ contain one
      factor of 3. This means the LCM will have one factor of 3.

Step 3: LCM $= (2 \cdot 2 \cdot 2)(5)(3)$ (3 factors of 2, 1 factor of 5,
        $= (8)(5)(3)$ and one factor of 3)
        $= 120$

In other words, 120 is the smallest number that both 30 and 24 can
divide into.

The key step in the above example is to determine which expression
contains the most of each factor. This same idea applies when the factors
are variables.

•Example 2:

Find the LCM of $2x^2 + 6x$ and $4x + 12$.

   Step 1: Factor each polynomial.

               $2x^2 + 6x$ and $4x + 12$
             $= 2x(x + 3)$ $= 4(x + 3)$
                                                 $= 2 \cdot 2(x + 3)$

   Step 2.   a. Which expression contains the most 2s?
                  Answer: $2 \cdot 2(x + 3)$ contains two factors of 2.

               b. Which expression contains the most xs?
                  Answer: The factor x appears most often in
                  $2x(x + 3)$. (one factor of x)

               c. Which expression contains the most $(x + 3)$ s?
                  Answer: Both $2x(x + 3)$ and $2 \cdot 2(x + 3)$ contain
                  one factor of $(x + 3)$.

   Step 3:   LCM $= (2 \cdot 2)(x)(x + 3)$ [2 factors of 2, 1 factor
               $= 4x(x + 3)$ of x, 1 factor of $(x + 3)$]

In the above example, it is proper to leave the LCM in its factored
form, $4x(x + 3)$.

•Example 3:

Find the LCM of $y^2 - 2y$ and $y^2 - 4y + 4$.

Step 1: $y^2 - 2y$ and $y^2 - 4y + 4$
           $= y(y - 2)$ $= (y - 2)(y - 2)$

Step 2:   a. Which expression contains the most $(y - 2)$ s?
             Answer: $(y - 2)(y - 2)$ contains two factors of $(y - 2)$.
            b. Which expression contains the most y factors?
             Answer: $y(y - 2)$ contains one factor of y.

Step 3:   LCM $= y(y - 2)(y - 2)$

Rational Expressions, Exercise 3. Find the LCM by the above method.

1. 18 and 30
2. 90 and 12
3. $x \cdot x$ and $x \cdot y$ Hint: Step 2
4. $x(x + 2)$ and $(x + 1)(x + 2)$
5. $x^2 + 2x$ and $x^2 + 3x + 2$ Hint: Example 3.
6. $y^2 - 3y$ and $y^2 + y - 12$
7. $y^2 - 16$ and $y^2 + y - 12$
8. $x^2 - y^2$ and $x^2 + 3xy + 2y^2$

9. $2x$ and 4
10. 9 and $3y$
11. 7 and 5
12. x and y
13. x and $(x + 2)$
14. xy and $y(x + 2)$
15. xy and y

Adding and Subtracting Rational Expressions

Remember from arithmetic that to add two fractions, you need a common denominator. The same is true for adding and subtracting rational expressions.

To Get a Common Denominator
Step 1: Factor each denominator.
Step 2: Identify factors that are present in one denonimator but absent (or missing) in the other.
Step 3: Multiply numerator and denominator by the missing factor.

•Example 1:

Add $\frac{7}{6}$ and $\frac{4}{15}$.

Step 1: $\quad \frac{7}{2\cdot 3} + \frac{4}{3\cdot 5}$

Step 2: 5 is present in the second denominator but absent in the first. 2 is present in the first denominator but absent in the second.

Step 3: Multiply numerator and denominator by the missing factors.

$$\frac{7}{2\cdot 3}\cdot\frac{5}{5} + \frac{4}{3\cdot 5}\cdot\frac{2}{2}$$

Now both denominators have all three factors. In other words, we have made a common denominator.

$$\frac{7\cdot 5}{2\cdot 3\cdot 5} + \frac{4\cdot 2}{3\cdot 5\cdot 2} = \frac{35}{30} + \frac{8}{30} = \frac{43}{30}$$

In the above example, it is legal to multiply by $\frac{5}{5}$ because $\frac{5}{5} = 1$. Multiplying by 1 never changes the value of a fraction.

•Example 2:

Add $\frac{2}{x+3} + \frac{6}{x^2+3x}$.

Step 1: $\quad \frac{2}{x+3} + \frac{6}{x(x+3)}$ $\qquad$ Factor out x.

Step 2: $\quad$ x + 3 is present in both denominators, however, the factor x is missing in the first.

Step 3. $\quad \frac{2}{(x+3)}\cdot\frac{x}{x} + \frac{6}{x(x+3)}$

Now denominators have the same factors, namely x and (x + 3). Therefore we can add the numerators together over the common denominators.

$$\frac{2x}{x(x+3)} + \frac{6}{x(x+3)} = \frac{2x+6}{x(x+3)}$$

We would be finished except that the fraction can be reduced as in the first section of this chapter.

$$\frac{2x+6}{x(x+3)} = \frac{2(x+3)}{x(x+3)} = \frac{2\cancel{(x+3)}}{x\cancel{(x+3)}} = \frac{2}{x} \quad \text{Finished.}$$

Example 2 suggests the following steps.

To Add Rational Expressions
Step 1: Factor denominators, then find a common denominator as described before.
Step 2: Place both numerators together over a common denominator and collect like terms.
Step 3: Reduce the new fraction, if possible.

•Example 3:

Add $\dfrac{x}{3x + 6} + \dfrac{x + 4}{x^2 + 7x + 10}$.

Step 1: $\dfrac{x}{3(x + 2)} + \dfrac{x + 4}{(x + 2)(x + 5)}$

The first denominator is missing the factor $(x + 5)$. The second denominator is missing the factor 3.

$\dfrac{x}{3(x + 2)} \cdot \dfrac{(x + 5)}{(x + 5)} + \dfrac{(x + 4)}{(x + 2)(x + 5)} \cdot \dfrac{3}{3}$

Now we have a common denominator, $3(x + 2)(x + 5)$.

Step 2: $\dfrac{x(x + 5) + (x + 4)3}{3(x + 2)(x + 5)} = \dfrac{x^2 + 5x + 3x + 12}{3(x + 2)(x + 5)}$ By distributing the x and the 3.

$= \dfrac{x^2 + 8x + 12}{3(x + 2)(x + 5)}$ Since $5x + 3x = 8x$.

Step 3: $\dfrac{x^2 + 8x + 12}{3(x + 2)(x + 5)} = \dfrac{(x + 6)(x + 2)}{3(x + 2)(x + 5)} = \dfrac{(x + 6)\cancel{(x + 2)}}{3\cancel{(x + 2)}(x + 5)}$

$= \dfrac{x + 6}{3(x + 5)}$ Finished.

Subtraction is Similar to Addition.

To Subtract Rational Expressions: Follow the identical process as for addition, except distribute the negative sign through the second numerator during Step 2.

44

•Example 4:

> Subtract $\dfrac{y+1}{y} - \dfrac{2-y}{y^2+6y}$.
>
> Step 1: $\qquad \dfrac{y+1}{y} - \dfrac{2-y}{y(y+6)}$
>
> The first denominator is missing the $(y+6)$ factor.
>
> $\qquad \dfrac{(y+1)}{y} \cdot \dfrac{(y+6)}{(y+6)} - \dfrac{2-y}{y(y+6)}$
>
> Step 2: $\qquad \dfrac{(y+1)(y+6) - (2-y)}{y(y+6)}$
>
> Multiply $(y+1)(y+6)$ and distribute the negative sign:
>
> $\dfrac{y^2 + 6y + 1y + 6 - 2 + y}{y(y+6)} = \dfrac{y^2 + 8y + 4}{y(y+6)}$ Finished.

•Example 5:

> Subtract $\dfrac{7}{x+2} - \dfrac{3}{x}$.
>
> Step 1: Both denominators are already factored.
> The factors $x+2$ and x look similar, but they are different factors. Indeed, the first denominator is missing the factor x. The second denominator is missing the factor $x+2$. Multiply each fraction by the missing factors to get:
>
> $\qquad \dfrac{7}{(x+2)} \cdot \dfrac{x}{x} - \dfrac{3}{x} \cdot \dfrac{(x+2)}{(x+2)}$
>
> Step 2: $\dfrac{7x - 3(x+2)}{x(x+2)} = \dfrac{7x - 3x - 6}{x(x+2)}$ By distributing -3.
>
> $= \dfrac{4x - 6}{x(x+2)}$ Finished.

Rational Expressions, Exercise 4.

Add

1. $\dfrac{7}{x} + \dfrac{2x-3}{x}$
Already common x.

2. $\dfrac{5y}{y+1} + \dfrac{5}{y+1}$
Hint: Reduce afterwards.

Subtract

7. $\dfrac{7}{x} - \dfrac{2x-3}{x}$

8. $\dfrac{4y}{y-2} - \dfrac{8}{y-2}$
Hint: Reduce afterwards.

3. $\dfrac{x}{2x-6} + \dfrac{5}{x-3}$

Hint: First factor 2x - 6.

4. $\dfrac{3}{y+2} + \dfrac{y}{y^2+2y}$

Hint: Example 2.

5. $\dfrac{1}{x+3} + \dfrac{2}{x^2+7x+12}$

Hint: Example 3.

6. $\dfrac{2}{x-2} + \dfrac{2}{x^2-5x+6}$

All Steps required.

9. $\dfrac{x}{x-3} - \dfrac{x+3}{2x-6}$

Hint: Factor 2x - 6 first.

10. $\dfrac{x-1}{x} - \dfrac{3}{x^2+2x}$

Hint: Example 4.

11. $\dfrac{5}{y-2} - \dfrac{3}{y}$

Hint: Example 5.

12. $\dfrac{1}{x} - \dfrac{1}{y}$

Factors Inside Fractions

Often in math or science, the result of some calculation or formula happens to be a fraction that contains another fraction. When this happens we would like to be able to eliminate the fraction in order to make the result less complicated.

To Eliminate Fractions Inside of a Fraction
Step 1: Find the LCM of **all** the denominators.
Step 2: Multiply the numerator and denominator of the **main** fraction by $\dfrac{\text{LCM}}{1}$.

•Example 1:

Simplify $\dfrac{\frac{1}{6} + \frac{7}{2}}{\frac{1}{3}}$.

Step 1: The LCM of 6, 2 and 3 is 6. $6 = \dfrac{6}{1}$

Step 2: $\dfrac{\left(\frac{1}{6} + \frac{7}{2}\right) \cdot \frac{6}{1}}{\left(\frac{1}{3}\right)\frac{6}{1}} = \dfrac{\frac{1}{6} \cdot \frac{6}{1} + \frac{7}{2} \cdot \frac{6}{1}}{\frac{1}{3} \cdot \frac{6}{1}}$

$\dfrac{6}{1}$ distributes to each fraction. Then denominators cancel.

$\dfrac{\frac{1}{6} \cdot \frac{6}{1} + \frac{7}{2} \cdot \frac{6^3}{1}}{\frac{1}{3} \cdot \frac{6^2}{1}} = \dfrac{1 + 21}{2} = \dfrac{22}{2} = 11$

•Example 2

Simplify $\dfrac{\dfrac{1}{x} - \dfrac{1}{2x}}{\dfrac{1}{2}}$.

Step 1: The LCM of x, 2x, and 2 is 2x.

Step 2: $\dfrac{\dfrac{1}{x} \cdot \dfrac{2x}{1} - \dfrac{1}{2x} \cdot \dfrac{2x}{1}}{\dfrac{1}{2} \cdot \dfrac{2x}{1}} = \dfrac{\dfrac{1}{\cancel{x}} \cdot \dfrac{\cancel{2x}}{1} - \dfrac{1}{\cancel{2x}} \cdot \dfrac{\cancel{2x}}{1}}{\dfrac{1}{\cancel{2}} \cdot \dfrac{\cancel{2}x}{1}}$

$= \dfrac{2 - 1}{x} = \dfrac{1}{x}$

•Example 3:

Simplify $\dfrac{\dfrac{3}{2y - 10}}{\dfrac{1}{2} + \dfrac{1}{y - 5}}$.

Solution: Factor 2y - 10 first to get: $\dfrac{\dfrac{3}{2(y - 5)}}{\dfrac{1}{2} + \dfrac{1}{y - 5}}$.

LCM = 2(y - 5)

$\dfrac{\dfrac{3}{2(y-5)} \cdot \dfrac{2(y-5)}{1}}{\dfrac{1}{2} \cdot \dfrac{2(y-5)}{1} + \dfrac{1}{y-5} \cdot \dfrac{2(y-5)}{1}} = \dfrac{\dfrac{3}{\cancel{2(y-5)}} \cdot \dfrac{\cancel{2(y-5)}}{1}}{\dfrac{1}{\cancel{2}} \cdot \dfrac{\cancel{2}(y-5)}{1} + \dfrac{1}{\cancel{y-5}} \cdot \dfrac{2\cancel{(y-5)}}{1}}$

$= \dfrac{3}{y - 5 + 2} = \dfrac{3}{y - 3}$ Finished.

Rational Expressions, Exercise 5.

1. $\dfrac{\dfrac{1}{2} + \dfrac{1}{3}}{\dfrac{1}{6}}$

5. $\dfrac{\dfrac{1}{x} + \dfrac{1}{x(x+2)}}{\dfrac{1}{x+2}}$ Hint: LCM = x(x +2)

2. $\dfrac{\dfrac{3}{4} - \dfrac{1}{6}}{\dfrac{5}{3}}$

6. $\dfrac{\dfrac{7}{x} - \dfrac{2}{x+3}}{\dfrac{1}{x^2 + 3x}}$ Hint: Factor $x^2 + 3x$ first.

3. $\dfrac{\dfrac{1}{x} + \dfrac{1}{y}}{\dfrac{1}{xy}}$ Hint: LCM = xy

7. $\dfrac{\dfrac{1}{x^2 - y^2} + \dfrac{y}{x+y}}{\dfrac{1}{x-y}}$ Hint: $x^2 - y^2 = (x + y)(x - y)$.

4. $\dfrac{\dfrac{1}{xy} - \dfrac{2}{x}}{\dfrac{3}{y}}$

Fractional Equations

In this section you will learn how to solve equations containing fractions. As in the previous section, this process involves using the LCM to eliminate the fractions. This process is called **clearing the fractions**.

•Example 1:

Solve $\dfrac{1}{2} - \dfrac{3}{x} = \dfrac{9}{2x}$.

Solution: The denominators are 2, x, and 2x. The LCM of all the denominators is 2x. Multiply each term on both sides of the equal sign by $\dfrac{2x}{1}$.

$\dfrac{2x}{1} \cdot \dfrac{1}{2} - \dfrac{3}{x} \cdot \dfrac{2x}{1} = \dfrac{9}{2x} \cdot \dfrac{2x}{1}$

$\dfrac{2x}{1} \cdot \dfrac{1}{\cancel{2}} - \dfrac{3}{\cancel{x}} \cdot \dfrac{\cancel{2x}}{1} = \dfrac{9}{\cancel{2x}} \cdot \dfrac{\cancel{2x}}{1}$ Cancel denominators.

$x \cdot 1 - 3 \cdot 2 = 9$
$x - 6 = 9$
$x = 15$

•Example 2:

Solve $\dfrac{4}{y} + \dfrac{1}{y-2} = \dfrac{7}{y^2-2y}$.

Solution: First factor the denominator $y^2 - 2y$.

$$\frac{4}{y} + \frac{1}{y-2} = \frac{7}{y(y-2)}$$

The LCM of the denominators is $y(y-2)$.
Multiply both sides by $y(y-2)$ over 1.

$$\frac{y(y-2)}{1} \cdot \frac{4}{y} + \frac{y(y-2)}{1} \cdot \frac{1}{y-2} = \frac{7}{y(y-2)} \cdot \frac{y(y-2)}{1}$$

$$\frac{\cancel{y}(y-2)}{1} \cdot \frac{4}{\cancel{y}} + \frac{y\cancel{(y-2)}}{1} \cdot \frac{1}{\cancel{y-2}} = \frac{7}{\cancel{y(y-2)}} \cdot \frac{\cancel{y(y-2)}}{1}$$

$$(y-2) \cdot 4 + y = 7$$
$$4y - 8 + y = 7 \qquad \text{Distribute the 4.}$$
$$5y = 15 \qquad \text{Collect like terms.}$$
$$y = 3$$

Sometimes, as in the next example, the fractional equation has no solution.

•Example 3:

Solve $10 + \dfrac{5}{x-2} = \dfrac{5}{x-2}$.

Solution: The LCM $= x - 2$. Multiply both sides by $x - 2$.

$$\frac{10}{1} \cdot \frac{(x-2)}{1} + \frac{5}{x-2} \cdot \frac{x-2}{1} = \frac{5}{x-2} \cdot \frac{x-2}{1}$$

$$10x - 20 + 5 = 5 \quad \text{Distribute 10, cancel } (x-2)\text{s.}$$
$$10x - 15 = 5$$
$$10x = 20$$
$$x = 2 \quad \text{Unfortunately, } x = 2 \text{ is}$$

not a proper solution, because replacing x with 2 in the
original equation causes both denominators to be zero:

$$10 + \frac{5}{x-2} = \frac{5}{x-2}$$

$$10 + \frac{5}{2-2} = \frac{5}{2-2} \qquad \text{Replacing x by 2.}$$

$$10 + \frac{5}{0} = \frac{5}{0}$$

It is **not** legal to have zero in the denominator of any fraction. Therefore, x = 2 is **not** a solution.

Hence, this example has no solutions. In other words, there are no values of x that will make the equation true.

The example below shows when the zero product rule can be used to finish a fractional equation problem.

•Example 4:

Solve $\frac{x}{6} - \frac{2}{3} = \frac{7}{2x}$.

Solution: The LCM = 6x.

$$\frac{x}{6} \cdot \frac{6x}{1} - \frac{2}{3} \cdot \frac{6x}{1} = \frac{7}{2x} \cdot \frac{6x}{1}$$

$$\frac{x}{\cancel{6}} \cdot \frac{\cancel{6}x}{1} - \frac{2}{\cancel{3}} \cdot \frac{\overset{2}{\cancel{6}}x}{1} = \frac{7}{\cancel{2x}} \cdot \frac{\overset{3}{\cancel{6x}}}{1}$$

$$x \cdot x - 2 \cdot 2 \cdot x = 7 \cdot 3$$

$x^2 - 4x = 21$ This is a **second degree** equation, meaning that the highest power on x is 2. To solve this equation, we must get zero on one side, factor, and apply the zero product rule.

$$x^2 - 4x - 21 = 0 \qquad \text{Subtract 21}$$

$$(x - 7)(x + 3) = 0 \qquad \text{Factor.}$$

$$x - 7 = 0 \quad \text{or} \quad x + 3 = 0 \qquad \text{Zero Product Rule.}$$

$$x = 7 \quad \text{or} \qquad x = -3$$

Examples 1 through 4 suggest the following steps.

To Solve a Fractional Equation:
Step 1: Factor any denominators which are not already factored (as in Example 2).
Step 2: Multiply every term by $\dfrac{LCM}{1}$ and cancel.
Step 3: If x^2 appears use the zero product rule (Example 4)
Step 4: Check the solutions to be sure they do not cause any denominator to be zero.

Rational Expressions, Exercise 6. Solve.

1. $\dfrac{5}{x} + \dfrac{2}{x} = \dfrac{7}{1}$
Hint: LCM = x.

6. $\dfrac{1}{x + 3} + \dfrac{x}{x^2 + 4x + 3} = \dfrac{1}{(x + 1)}$
Hint: Step 1.

2. $\dfrac{5}{2} - \dfrac{1}{x} = \dfrac{3}{2x}$
Hint: Example 1.

7. $\dfrac{3}{2} + \dfrac{7}{x} = \dfrac{7}{x}$
Hint: Check back of the book.

3. $\dfrac{7}{10} + \dfrac{1}{2} = \dfrac{6}{5x}$
Hint: LCM = 10x

8.* $\dfrac{1}{x(x - 2)} + \dfrac{x}{x - 2} = \dfrac{10}{x}$

4. $\dfrac{3}{(x - 1)} + \dfrac{1}{x(x - 1)} = \dfrac{2}{x}$
Hint: Example 2.

9.* $\dfrac{x}{10} - \dfrac{1}{2} = \dfrac{-3}{5x}$

5. $\dfrac{5}{x^2 + 4x} = \dfrac{3}{x} - \dfrac{2}{x+4}$
Hint: first factor $x^2 + 4x$.

10. $\dfrac{5}{x - 1} = \dfrac{5}{x}$
Hint: LCM = x(x - 1).

11. In problem 4, check that the solution does not make any denominator zero.

*Problems 8 and 9 use the zero product rule as in Example 4.

51

Review

Chapter 1

1. Plot these ordered pairs:

     A. (2, 4)   B. (-3, 1)   C. (3, -1)   D. (-5, -5)   E. (0, -3)

2. Which are solutions to $3x - 4y = 1$?

     A. (0, 2)   B. (-1, -1)   C. (7, 5)

3. A. Graph $y = 2x - 3$ by the point plotting method.

   B. Graph $-3x + 2y = 6$ by using zeros.

   C. Graph $y = -2$.

4. Use the slope formula to find the slope of a line which passes through (1, 1) and (5, -2).

5. Determine the slope and y-intercept of each:

   A. $y = 9x + 1$ B. $y = -\frac{1}{2}x - 3$ C. $2x + y = 3$

6. Use the slope intercept method to graph the lines in Parts A and B of Exercise 3: that is, $y = 2x - 3$ and $-3x + 2y = 6$.

7. Use $y = mx + b$ to find the equation of a line which passes through (2, -1) with a slope of $-\frac{1}{2}$.

Chapter 2

Determine whether or not the ordered pairs are solutions to the system.

1. $5x - 4y = 2$
   $-2x + y = 1$

   A. (2, 2)   B. (-2, -3)

2. $x = y + 2$
   $3x - 3y = 6$

   A. (2, 0)   B. (-5, -3)

3. Solve: $5x - 4y = 2$
            $-2x + y = 1$
   Using (a.) substitution and (b.) addition method.

4. Solve: $x = y + 2$
            $3x - 3y = 6$
   By (a.) graphing, (b.) substitution, and
          (c.) addition method.

Chapter 3

1. Reduce;

    A. $\dfrac{x}{x \cdot y}$
    B. $\dfrac{x}{x^2 + 3x}$
    C. $\dfrac{x^2 - 25}{x + 5}$
    D. $\dfrac{x^2 - 5x + 6}{x^2 - 2x}$

2. Multiply or divide:

    A. $\dfrac{x}{x + 1} \cdot \dfrac{5x + 5}{x^2 + x}$
    B. $\dfrac{x^2 + 7x + 10}{x^2 + 2x + 1} \div \dfrac{(x + 5)}{(x + 1)}$

3. Add or subtract:

    A. $\dfrac{5}{x + 1} + \dfrac{5}{x^2 + x}$
    B. $\dfrac{2}{x^2 - 4} + \dfrac{1}{x + 2}$
    C. $\dfrac{1}{x} - \dfrac{1}{y}$

4. Simplify:

    A. $\dfrac{\frac{1}{x} + \frac{1}{y}}{\frac{1}{xy}}$
    B. $\dfrac{\frac{1}{2x - 10} + \frac{3}{x - 5}}{\frac{1}{2}}$

5. Solve:

    A. $\dfrac{1}{x} + \dfrac{1}{3} = \dfrac{8}{x}$
    B. $\dfrac{4}{x - 3} + \dfrac{1}{x} = \dfrac{2}{x^2 - 3x}$

Radical Expressions

A radical expression is an expression that contains the symbol $\sqrt{}$. For example:

$$\sqrt{4} \qquad \sqrt{x} \qquad 7\sqrt{5y} \qquad -x\sqrt{Y+3}$$

The symbol $\sqrt{}$ is called a radical sign.

The part of the expression that is inside the radical sign is called the **radicand**. For instance in $-x\sqrt{Y+3}$, the radicand is $Y+3$. Also in $7\sqrt{5y}$ the radicand is 5y.

Radical expressions are used in sciences like physics, engineering, and surveying. They are also used in mathematics like trigonometry and algebra. For example, we will use radical expressions in Chapter 5 to solve second degree equations.

In order to understand radical expressions, we need a concept called the **square roots of a number**.

The Square Roots of a Number are numbers which, when squared, are equal to the original number.

•Example 1:

> Find the square roots of 49.
>
> Solution: What number squared is 49?
> $7^2 = 7 \cdot 7 = 49$, so 7 is a square root of 49.
>
> Likewise, the number -7 is also a square root of 49.
> Since $(7)^2 = (-7)(-7) = 49$.
>
> Therefore the square roots of 49 are 7 and -7.

As in the above example, every positive number has two square roots: one positive square root and one negative square root.

The positive square root is called the principal root. Example 1 shows that 7 is the principal square root of 49 and -7 is the negative square root of 49.

The radical sign, $\sqrt{}$, is used to denote the principal square root. The radical sign with a negative in front, $-\sqrt{}$, is used to denote the negative square root. For instance, in Example 1, we could write $\sqrt{49} = 7$ and $-\sqrt{49} = -7$.

•Example 2:

> Calculate each square root: $\sqrt{25}$ and $-\sqrt{25}$.
>
> Solution: What number squared equals 25?
>
> $5^2 = 25$ and $(-5)^2 = (-5)(-5) = 25$.
> Therefore, the principal root is 5 and the negative root is -5.
>
> That is, $\sqrt{25} = 5$ and $-\sqrt{25} = -5$.

•Example 3:

> Calculate $\sqrt{0}$.
>
> Solution: What number squared equals 0 ?
>
> $0^2 = 0 \cdot 0 = 0$. Hence $\sqrt{0} = 0$.

Examples 1 and 2 demonstrate that positive numbers have two square roots. Example 3 shows that zero has a square root.

Negative numbers, on the other hand, do not have square roots. This is because you can never get a negative result when squaring numbers. For instance, $(-4)^2 = +16$, $(-5)^2 = +25$, $(-3)^2 = +9$, and $3^2 = +9$. Therefore, any number like -9 has no chance of having square roots, because the square of any non-zero number is always positive.

Since negative numbers do not have square roots it is meaningless to place a negative inside of a radical sign. In other words, $\sqrt{-9}$ has no meaning.

•Example 4:

> Calculate the following. State if an expression has no meaning.
>
> a. $\sqrt{9}$ c. $\sqrt{-64}$ e. $\sqrt{100}$ g. $\sqrt{-4}$
>
> b. $\sqrt{-9}$ d. $-\sqrt{64}$ f. $\sqrt{-100}$ h. $-\sqrt{4}$
>
> Solutions:
>
> a. 3 c. no meaning e. 10 g. no meaning
> b. no meaning d. -8 f. no meaning h. -2

Since $\sqrt{-9}$, $\sqrt{-64}$, $\sqrt{-100}$, and $\sqrt{-4}$ have no meaning, it is also true that $\sqrt{x}$ has no meaning if x represents a negative number. Therefore, in this book, we will asume that all variables inside radical signs represent non-negative numbers. For instance, in the expressions $\sqrt{xy}$, we assume that x and y are non-negative.

Radical Expressions, Exercise 1.

1. In the expression $8\sqrt{3y}$, the $\sqrt{}$ is called_____ and 3y is called the_____.

2. In $7\sqrt{xy}$, the radicand is _____

3. Fill in the missing numbers. That is, what number squared equals the number to the right of the equal sign? If there is no number that will work, state so.

a. $(7)^2 = 49$ c. $(\ )^2 = 1$ e. $(\ )^2 = 0$ g. $(\ )^2 = -25$

b. $(\ )^2 = 64$ d. $(\ )^2 = -4$ f. $(\ )^2 = 4$ h. $(\ )^2 = \frac{1}{25}$

4. Find the principal and negative square roots of each number. If a number has no square root, state so.

a. 81   b. 100   c. -49   d. 0

5. Calculate each. If an expression has no meaning, state so.

a. $\sqrt{9}$ d. $\sqrt{-16}$ g. $-\sqrt{1}$ j. $-\sqrt{100}$

b. $-\sqrt{9}$ e. $\sqrt{16}$ h. $\sqrt{-1}$ k. $\sqrt{0}$

c. $\sqrt{-9}$ f. $-\sqrt{16}$ i. $\sqrt{1}$ l. $\sqrt{36}$

Perfect Square Numbers and Radicands

Any number whose square root is a whole number is called a perfect square. For example, 4, 9, and 16 are perfect squares, since: $\sqrt{4} = 2$, and $\sqrt{9} = 3$, and $\sqrt{16} = 4$. We can make a list of perfect squares by squaring each whole number.

Whole numbers:	0	1	2	3	4	5	6	etc....
Square each:	0^2	1^2	2^2	3^2	4^2	5^2	6^2	etc....
Perfect Squares:	0	1	4	9	16	25	36	etc....

Perfect square numbers are important, because their square roots can be quickly calculated. In the following section, perfect squares will be used to simplify more complicated radical expressions like $\sqrt{12}$ or $\sqrt{18}$.

Another concept that will be useful to us in this chapter is called **perfect square radicands**.

A perfect square radicand is an expression inside the radical sign which consists of either two identical factors or a single factor raised to the power of 2. For example, $\sqrt{5 \cdot 5}$, $\sqrt{x \cdot x}$, $\sqrt{3^2}$, and $\sqrt{y^2}$ are radical expressions that contain perfect square radicands. Perfect square radicands are important, because they can be immediately simplified as explained below.

Consider the radical expressions $\sqrt{2 \cdot 2}$, $\sqrt{3 \cdot 3}$, $\sqrt{4 \cdot 4}$, and $\sqrt{5 \cdot 5}$. Let's calculate each of these and look for a pattern:

$$\sqrt{2 \cdot 2} = \sqrt{4} = 2$$
$$\sqrt{3 \cdot 3} = \sqrt{9} = 3$$
$$\sqrt{4 \cdot 4} = \sqrt{16} = 4$$
$$\sqrt{5 \cdot 5} = \sqrt{25} = 5$$

The above examples show that when a number is multiplied by itself under the radical sign, the square root will always be that same number. In other words, $\sqrt{2 \cdot 2} = 2$, $\sqrt{3 \cdot 3} = 3$, $\sqrt{4 \cdot 4} = 4$, and $\sqrt{5 \cdot 5} = 5$.

This pattern leads to the belief that $\sqrt{x \cdot x} = x$, provided that x represents a non-negative number.

•Example 1:

> Calculate the following. Assume all variables represent non-negative numbers.
>
> a. $\sqrt{6 \cdot 6}$ d. $\sqrt{c^2}$ g. $\sqrt{A^2}$
> b. $\sqrt{x \cdot x}$ e. $\sqrt{(1.3)(1.3)}$ h. $\sqrt{(xy)(xy)}$
> c. $\sqrt{8^2}$ f. $\sqrt{(51)^2}$ i. $\sqrt{(ab)^2}$
>
> Solutions: All radicands are perfect squares.
>
> a. $\sqrt{6 \cdot 6} = 6$ d. $\sqrt{c^2} = \sqrt{c \cdot c} = c$ g. $\sqrt{A^2} = A$
> b. $\sqrt{x \cdot x} = x$ e. $\sqrt{(1.3)(1.3)} = 1.3$ h. $\sqrt{(xy)(xy)} = (xy)$
> c. $\sqrt{8^2} = \sqrt{8 \cdot 8} = 8$ f. $\sqrt{51^2} = 51$ g. $\sqrt{(ab)^2} = (ab)$

Radical Expressions, Exercise 2.

1. In this section, it was shown that the first seven perfect square numbers are 0, 1, 4, 9, 16, 25, and 36. What are the next seven perfect square numbers after 36?

2. Calculate the following. Assume all variables represent non-negative numbers.

a. $\sqrt{5 \cdot 5}$　　　　e. $\sqrt{c \cdot c}$　　　　i. $\sqrt{(x + 3)^2}$

b. $\sqrt{5^2}$　　　　f. $\sqrt{(127)^2}$　　　　j. $\sqrt{(xy)^2}$

c. $\sqrt{9^2}$　　　　g. $\sqrt{(0.75)(0.75)}$　　　　k. $\sqrt{(y + 2)(y + 2)}$

d. $\sqrt{C^2}$　　　　h. $\sqrt{x^2}$　　　　l. $\sqrt{(2x + 7)^2}$

3. The radicands in Problem 2 are called _____ radicands.

Simplifying Radical Expressions

Thus far, we have considered only radicals containing perfect square numbers or perfect square expressions. In this section, you will learn how to simplify square roots of non-perfect square numbers. In order to simplify these types of radical expressions, we need a concept called the **Product Rule**

The Product Rule: It is legal to split a radical expression into smaller pieces provided that the factors are connected to each other by multiplication. In other words: $\sqrt{A \cdot B} = \sqrt{A} \cdot \sqrt{B}$.

•Example 1:

$$\sqrt{4 \cdot 9} = \sqrt{4} \cdot \sqrt{9}$$

We can work both sides of the equal sign to show this is true.

Left side　　　　　　　　　　　　Right side

$\sqrt{4 \cdot 9}$　　　　　　　　　　　$\sqrt{4} \cdot \sqrt{9}$
$\sqrt{36}$　　　　　　　　　　　　　$2 \cdot 3$
6　　　　　　　　　　　　　　　　6

Both sides are 6. Hence $\sqrt{4 \cdot 9} = \sqrt{4} \cdot \sqrt{9}$.

The next example shows how the Product Rule enables us to simplify radical expressions. We simplify radicals because we want them to look less complicated.

58

•Example 2

Simplify $\sqrt{44}$.

$$
\begin{aligned}
\text{Solution:} \quad \sqrt{44} &= \sqrt{4 \cdot 11} \qquad &&\text{Factor 44 first.} \\
&= \sqrt{4} \cdot \sqrt{11} &&\text{Product Rule.} \\
&= 2 \cdot \sqrt{11} &&\text{Since } \sqrt{4} = 2. \\
&= 2 \sqrt{11}
\end{aligned}
$$

The next example is done two ways to show that the same answer is obtained regardless of how the radicand is factored.

•Example 3:

Simplify $\sqrt{360}$.

First Way: Our goal is to factor 360 in such a way that one of the factors is a perfect square number.

Ask yourself: What perfect square number divides evenly into 360? 9 is a perfect square number that divides 360, so:

$$
\begin{aligned}
\sqrt{360} &= \sqrt{9 \cdot 40} \\
&= \sqrt{9} \cdot \sqrt{40} \qquad &&\text{Product Rule.} \\
&= 3 \cdot \sqrt{40}
\end{aligned}
$$

We are not finished, however, until we simplify $\sqrt{40}$. What perfect square number divides into 40? 4 is a perfect square that divides 40, so:

$$
\begin{aligned}
3\sqrt{40} &= 3 \cdot \sqrt{4 \cdot 10} \\
&= 3 \cdot \sqrt{4} \cdot \sqrt{10} \qquad &&\text{Product Rule.} \\
&= 3 \cdot 2 \cdot \sqrt{10} \\
&= 6 \cdot \sqrt{10}
\end{aligned}
$$

We have finished, since there are no perfect square numbers that divide into 10.

Second Way: This process can be greatly shortened by finding the largest perfect square that divides 360. The largest square that divides 360 is 36.

$$
\begin{aligned}
\sqrt{360} &= \sqrt{36 \cdot 10} \\
&= \sqrt{36} \cdot \sqrt{10} \qquad &&\text{Product Rule} \\
&= 6 \sqrt{10} &&\text{Same answer.}
\end{aligned}
$$

The second way is much shorter, but either way works.

The previous examples have shown how to simplify $\sqrt{44}$ and $\sqrt{360}$ by finding perfect square numbers that divide into 44 and 360. Sometimes, however, it is difficult to find a perfect square number that divides into the radicand. For example, in a problem like $\sqrt{343}$, it may take awhile to discover that 49 is a perfect square that divides 343.

Fortunately, there is an alternative method which uses prime numbers instead of perfect square numbers.

•Example 4 An Alternative Method:

Simplify $\sqrt{360}$.

Solution: Begin by factoring 360 into prime numbers.

$$\sqrt{360} \quad = \sqrt{2 \cdot 180} = \sqrt{2 \cdot 2 \cdot 90} \quad = \sqrt{2 \cdot 2 \cdot 3 \cdot 30}$$

$$= \sqrt{2 \cdot 2 \cdot 3 \cdot 3 \cdot 2 \cdot 5}$$

Now use the Product Rule to separate out pairs of identical factors:

$$\sqrt{2 \cdot 2 \cdot 3 \cdot 3 \cdot 2 \cdot 5} = \sqrt{2 \cdot 2} \cdot \sqrt{3 \cdot 3} \cdot \sqrt{2 \cdot 5}$$

$$= 2 \cdot 3 \cdot \sqrt{2 \cdot 5} \text{ Because } \sqrt{2 \cdot 2} = 2, \text{ and } \sqrt{3 \cdot 3} = 3.$$

$$= 6 \sqrt{10} \text{ as in Example 3.}$$

•Example 5:

Simplify $\sqrt{343}$.

Solution: $\quad \sqrt{343} \quad$ $\begin{aligned} &= \sqrt{7 \cdot 7 \cdot 7} &&\text{Since } 7 \cdot 7 \cdot 7 = 343. \\ &= \sqrt{7 \cdot 7} \cdot \sqrt{7} &&\text{Product Rule.} \\ &= 7 \sqrt{7} &&\text{Because } \sqrt{7 \cdot 7} = 7. \\ &= 7\sqrt{7} \end{aligned}$

Radical Expressions, Exercise 3.

1. Verify each product equation by working each side separately as in Example 1.

    a. $\sqrt{4 \cdot 16} = \sqrt{4} \cdot \sqrt{16}$ b. $\sqrt{4 \cdot 25} = \sqrt{4} \sqrt{25}$

2. Use the Product Rule to fill in the missing factors:

a. $\sqrt{3 \cdot 4} = \sqrt{} \cdot \sqrt{}$ c. $\sqrt{4} \cdot \sqrt{11} = \sqrt{ \cdot }$

b. $\sqrt{x \cdot x} = \sqrt{} \cdot \sqrt{}$ d. $\sqrt{x^3} \cdot \sqrt{x} = \sqrt{ \cdot }$

3. Simplify by finding a perfect square factor as in Example 2 and 3.

a. $\sqrt{12}$ d. $\sqrt{20}$ g. $\sqrt{63}$ j. $\sqrt{72}$

b. $\sqrt{18}$ e. $\sqrt{45}$ h. $\sqrt{32}$ k. $\sqrt{600}$

c. $\sqrt{50}$ f. $\sqrt{8}$ i. $\sqrt{48}$ l. $\sqrt{108}$

4. Simplify by factoring each radicand into prime numbers as in Examples 4 and 5. Hint: It is legal to rearrange the prime factors when finding pairs of identical factors.

a. $\sqrt{8}$ d. $\sqrt{40}$ g. $\sqrt{54}$ j. $\sqrt{200}$

b. $\sqrt{28}$ e. $\sqrt{27}$ h. $\sqrt{32}$ k. $\sqrt{72}$

c. $\sqrt{75}$ f. $\sqrt{24}$ i. $\sqrt{60}$ l. $\sqrt{600}$

Variable Radicands

The square root of a variable expression is an expression which, when squared, equals the original expression.

•Example 1:

> Find $\sqrt{x^8}$.
>
> Solution: What expression equals x^8 when squared? Well, $(x^4)^2 = x^8$ by the Power Rule of Exponents.
>
> Therefore, $\sqrt{x^8} = x^4$.

Notice in Example 1 that the power in the answer, namely 4, is exactly half of the power in the radicand, 8. This observation is the essence of the **Even Exponent Rule**.

Even Exponent Rule: If a number or variable is raised to an even power, then the square root can be calculated by recopying the base and dividing the power by 2. In other words, $\sqrt{A^n} = A^{n/2}$, provided that n is even and A is non-negative.

•Example 2:

> Use the Even Exponent Rule to calculate:
>
> a. $\sqrt{y^{50}}$ b. $\sqrt{x^2\,y^{10}}$ c. $\sqrt{2^8\,(y+2)^4}$
>
>   Solutions: Divide all the powers by 2 to get:
>
>   a. y^{25} b. $x^1\,y^5$ or xy^5 c. $2^4(y+2)^2$

If a radical contains an odd power, then rewrite the expression using the closest even power.

•Example 3:

> Find (a.) $\sqrt{x^7}$ (b.) $\sqrt{2y^3}$
>
>   Solution: (a.) $\sqrt{x^7}$ $= \sqrt{x^6 \cdot x^1}$ (b.) $\sqrt{2y^3}$ $= \sqrt{2y^2 y}$
> $= \sqrt{x^6} \cdot \sqrt{x}$ Product Rule $= \sqrt{y^2}\,\sqrt{2y}$
> $= x^3\,\sqrt{x}$ $= y\,\sqrt{2y}$

Next we learn how to simplify expressions which contain numbers and variables.

To Simplify in General
Step 1: Factor the radicand.
Step 2: If a factor has a square root, then calculate its square root and write it down in front of the radical sign.
Step 3: Factors that are outside of the radical sign are multiplied together, as are factors inside the radical sign.

•Example 4:

> Simplify $\sqrt{45x^5}$.
>
> Step 1: $\sqrt{45x^5}$ $=$ $\sqrt{9 \cdot 5\, x^4 \cdot x}$
>
> Step 2:   The factors 9 and x^4 have square roots and can be removed from the radical sign.
>
> $\sqrt{9 \cdot 5 \cdot x^4 \cdot x}$
>
> $= 3 \cdot x^2 \sqrt{5 \cdot x}$ Since $\sqrt{9} = 3$ and $\sqrt{x^4} = x^2$.
>
> Step 3: $= 3x^2\,\sqrt{5x}$

•Example 5:

Simplify $\sqrt{135x^3}$.

Solution: Since 135 is a large number, we choose to factor 135 into prime numbers.

$$135 = 5 \cdot 27 = 5 \cdot 3 \cdot 3 \cdot 3$$

Step 1: $\sqrt{135\,x^3} = \sqrt{5 \cdot 3 \cdot 3 \cdot 3 \cdot x^2 \cdot x}$

Step 2: The x^2 factor has a square root and will therefore be removed from the radical. Also, a pair of 3s can be removed:

$$\sqrt{5 \cdot \mathbf{3 \cdot 3} \cdot 3 \cdot \mathbf{x^2} \cdot x} = 3x\sqrt{5 \cdot 3 \cdot x} \quad \text{Since } \sqrt{x^2} = x, \text{ and } \sqrt{3 \cdot 3} = 3.$$

Step 3: $3x\sqrt{15x}$ Finished.

Radical Expressions, Exercise 4.

1. Fill in the missing expressions. That is, what expression squared equals the expression to the right of the equal sign?

a. $(x^4)^2 = x^8$

b. $(\quad)^2 = y^6$

c. $(\quad)^2 = x^2$

d. $(\quad)^2 = 2^{10}$

e. $(\quad)^2 = x^6 y^8$

f. $(\quad)^2 = 2^{10} y^2$

2. Simplify each.

a. $\sqrt{x^6}$

b. $\sqrt{2^{10}}$

c. $\sqrt{y^{100}}$

d. $\sqrt{a^2 b^8}$

e. $\sqrt{y^7}$

f. $\sqrt{x^{11}}$

g. $\sqrt{3y^5}$

h. $\sqrt{x^3}$

3. Simplify. If a number is large, then follow Example 5.

a. $\sqrt{16x^5}$

b. $\sqrt{27x^4 y^2}$

c. $\sqrt{50ab^7}$

d. $\sqrt{200a^{11} b^2}$

e. $\sqrt{99x^4 y^4}$

f. $\sqrt{32xy}$

g. $\sqrt{75a^{21} b^2 c^2}$

h. $\sqrt{112xy^{24}}$

i. $\sqrt{375x^7 yz^{13}}$

Multiplying Radicals Together

The product rule enables us to multiply two radical expressions together. The result is then simplified as in the previous section.

•Example 1:

Multiply $\sqrt{6x} \cdot \sqrt{10x}$.

Solution: $\sqrt{6x} \cdot \sqrt{10x}$ $= \sqrt{6x \cdot 10x}$ By the product rule.

$= \sqrt{60\,x^2}$

Now simplify as in the previous section:

$\sqrt{60x^2} = \sqrt{4 \cdot 15x^2} = 2x\sqrt{15}$

•Example 2:

Multiply $\sqrt{21xy} \cdot \sqrt{28x}$.

Solution: $= \sqrt{21xy} \cdot \sqrt{28x}$

$= \sqrt{21 \cdot 28x^2y}$ Product Rule

$= \sqrt{3 \cdot 7 \cdot 2 \cdot 2 \cdot 7 \cdot x^2y}$ Since $21 = 3 \cdot 7$, $28 = 2 \cdot 2 \cdot 7$.

$= 2 \cdot 7 \cdot x\sqrt{3y}$ By removing a pair of 7s and 2s.

$= 14x\sqrt{3y}$

In Example 2, the numbers 21 and 28 were factored into primes rather than multiplied together. Multiplying them would result in a larger number that would have to be factored anyway.

•Example 3:

Multiply $2\sqrt{3ab}$ and $4\sqrt{15a}$.

Solution: $2\sqrt{3ab} \cdot 4\sqrt{15a}$

$= 2 \cdot 4 \cdot \sqrt{3ab} \cdot \sqrt{15a}$ Commutative Law of Multiplication

$= 8 \cdot \sqrt{3ab \cdot 15a}$ Product rule.

$= 8 \cdot \sqrt{3 \cdot 3 \cdot 5a^2 b}$ Because $15 = 3 \cdot 5$.

$= 8 \cdot 3 \cdot a\sqrt{5b}$ Removing a pair of 3s and a ^{2}.

$= 24a\sqrt{5b}$

Radical Expressions, Exercise 5. Multiply and simplify.

1. $\sqrt{2} \cdot \sqrt{6}$ 4. $\sqrt{y^5} \cdot \sqrt{y^3}$ 7. $2\sqrt{10} \cdot \sqrt{6}$

2. $\sqrt{3x} \cdot \sqrt{6}$ 5. $\sqrt{x} \cdot \sqrt{x}$ 8. $x\sqrt{10x} \cdot 7\sqrt{15x}$

3. $\sqrt{5} \cdot \sqrt{5}$ 6. $\sqrt{2a^2 b} \cdot \sqrt{4a^2}$ Hint: $7x\sqrt{10x} \cdot \sqrt{15x}$

In 9 through 17, factor the number into primes as in Example 2.

9. $\sqrt{14} \cdot \sqrt{21}$ 12. $\sqrt{55xy} \cdot \sqrt{77x}$ 15. $x\sqrt{x} \cdot 2\sqrt{x}$

10. $\sqrt{15x} \cdot \sqrt{21x}$ 13. $2\sqrt{7} \cdot \sqrt{7}$ 16. $a\sqrt{a^3} \cdot 5\sqrt{a^2}$

11. $\sqrt{35y^2} \cdot \sqrt{15y^4}$ 14. $\sqrt{33a^{11}b^{11}} \cdot \sqrt{3}$ 17. $\sqrt{20} \cdot \sqrt{20}$

Collecting Like Terms

Two radical expressions are called **like terms** if they have the same radical.

For example: $5\sqrt{3}$ and $7\sqrt{3}$ are like terms since they both have $\sqrt{3}$.

To Collect Like Terms: Add the coefficients together, followed by their common radical.

•Example 1:

> Collect like terms:
>   a. $2\sqrt{x} + 3\sqrt{x}$ b. $7\sqrt{2} - \sqrt{2}$
>
> Solution: a. $2\sqrt{x} + 3\sqrt{x} = 5\sqrt{x}$, because $2 + 3 = 5$.
>   b. $7\sqrt{2} - \sqrt{2} = 6\sqrt{2}$, because $7 - 1 = 6$.

•Example 2:

> Simplify $4\sqrt{7} + \sqrt{x} + 5\sqrt{7} - 3\sqrt{x}$.
>
> Solution: $9\sqrt{7} - 2\sqrt{x}$, since $4 + 5 = 9$ and $1 - 3 = -2$.

In Example 2, $9\sqrt{7} - 2\sqrt{x}$ cannot be combined since they are not like terms.

•Example 3:

Add $\sqrt{18} + \sqrt{2}$.

    Solution: $\sqrt{18}$ and $\sqrt{2}$ do not appear to be like terms. However, we can make them like each other by simplifying $\sqrt{18}$.

$$
\begin{aligned}
\sqrt{18} + \sqrt{2} &= \sqrt{9 \cdot 2} + \sqrt{2} \\
&= 3\sqrt{2} + \sqrt{2} \quad \text{Since } \sqrt{9} = 3. \\
&= 4\sqrt{2} \quad\quad\;\; \text{Combine like terms.}
\end{aligned}
$$

•Example 4:

Simplify $\sqrt{50} + 3\sqrt{4x} - \sqrt{12} + \sqrt{x}$.

    Solution: Begin by simplifying each term:

$$
\begin{aligned}
&\sqrt{50} + 3\sqrt{4x} - \sqrt{12} + \sqrt{x} \\
&= \sqrt{(25)\,2} + 3 \cdot 2\sqrt{x} - \sqrt{4 \cdot 3} + \sqrt{x} \quad\quad \text{Since } \sqrt{4x} = 2\sqrt{x}. \\
&= 5\sqrt{2} + 6\sqrt{x} - 2\sqrt{3} + \sqrt{x} \quad\quad\;\; \text{Since } \sqrt{25} = 5, \sqrt{4} = 2. \\
&= 3\sqrt{2} + 7\sqrt{x} \quad\quad\quad\quad\quad\quad\quad\;\; \text{Since } 5 - 2 = 3, 6 - 1 = 7.
\end{aligned}
$$

Rational Expressions, Exercise 6. Simplify.

1. $5\sqrt{y} - 3\sqrt{y}$

2. $10\sqrt{2} - \sqrt{2}$

3. $3\sqrt{3} + 3\sqrt{3}$

4. $10\sqrt{y} - 7\sqrt{y} + \sqrt{5}$

5. $\sqrt{18} + \sqrt{32}$ See Example 3 or 4

6. $\sqrt{28} - \sqrt{63}$

7. $\sqrt{8} - \sqrt{50} + \sqrt{12}$

8. $\sqrt{9x} + \sqrt{16x} - \sqrt{25x}$

9. $\sqrt{72} + \sqrt{50}$

10. $\sqrt{18xy} - \sqrt{8xy}$

11. $\sqrt{4a^2} + \sqrt{9a^2}$

12. $10\sqrt{y^5} - \sqrt{4y^5}$

Simplifying Radicals Containing Fractions

We simplify radicals to make them look less complicated. In earlier sections, the product rule was used to simplify. If the radical contains a fraction, we use the Quotient Rule to simplify.

Quotient Rule: The square root of a fraction can be found by taking the square root of the numerator and the square root of the denominator separately. In other words: $\sqrt{\dfrac{A}{B}} = \dfrac{\sqrt{A}}{\sqrt{B}}$ provided that A and B are non-negative.

•Example 1:

$$\sqrt{\frac{100}{25}} = \frac{\sqrt{100}}{\sqrt{25}}.$$ We can work both sides of the equation to show this is true:

Left Side   Right Side

$$\sqrt{\frac{100}{25}} = \sqrt{\frac{\cancel{100}}{\cancel{25}} \cdot \frac{4}{1}} = \sqrt{\frac{4}{1}} = \sqrt{4} = 2 \qquad \frac{\sqrt{100}}{\sqrt{25}} = \frac{10}{5} = \frac{2}{1} = 2$$

Both sides equal 2. Therefore $\sqrt{\frac{100}{25}} = \frac{\sqrt{100}}{\sqrt{25}}$.

Now we use the Quotient Rule to simplify.

•Example 2:

Simplify $\dfrac{\sqrt{18}}{\sqrt{8}}$.

Solution: Neither 18 nor 8 is a perfect square number, so we cannot calculate $\sqrt{18}$ or $\sqrt{8}$. Fortunately, the Quotient Rule allows us to rewrite the entire fraction under one radical sign.

$$\frac{\sqrt{18}}{\sqrt{8}} \qquad = \sqrt{\frac{18}{8}} \qquad \text{By Quotient Rule.}$$

$$= \sqrt{\frac{9}{4}} \qquad \text{By reducing } \frac{18}{8} = \frac{9}{4}.$$

$$= \frac{\sqrt{9}}{\sqrt{4}} \qquad \text{Quotient Rule again.}$$

$$= \frac{3}{2} \qquad \text{Since } \sqrt{9} = 3, \sqrt{4} = 2.$$

Example 2 suggests the following steps:
Step 1: Use the Quotient Rule to write the entire fraction under one radica sign.
Step 2: Reduce the fraction inside the radical sign.
Step 3: Use the Quotient Rule again to write the numerator in one radical and the denominator in another radical.

•Example 3:

Simplify $\dfrac{\sqrt{32x^3}}{\sqrt{50x}}$

Step 1: $\dfrac{\sqrt{32x^3}}{\sqrt{50x}} = \sqrt{\dfrac{32x^3}{50x}}$ Quotient Rule

Step 2: $= \sqrt{\dfrac{16x^2}{25}}$ Because $\dfrac{x^3}{x} = \dfrac{x^3}{x^1} = x^2$.

Step 3: $= \dfrac{\sqrt{16x^2}}{\sqrt{25}}$ Quotient Rule.

$= \dfrac{4x}{5}$

•Example 4:

Simplify $\dfrac{\sqrt{75y^5}}{\sqrt{4y^3}}$

Step 1 and 2: $\dfrac{\sqrt{75y^5}}{\sqrt{4y^3}} = \sqrt{\dfrac{75y^2}{4}}$ Since $\dfrac{y^5}{y^3} = y^2$.

Step 3: $= \dfrac{\sqrt{75y^2}}{\sqrt{4}} = \dfrac{\sqrt{25 \cdot 3y^2}}{2} = \dfrac{5y\sqrt{3}}{2}$

In Example 4, the final result contains a square root in its numerator. It is all right if the final result has a square root in its numerator.

In contrast, it is not considered proper to leave your answer with a square root in the denominator. When a fraction has a square root in the denominator, we need to get rid of it using a process called **rationalizing the denominator**.

To Rationalize the Denominator: Multiply the numerator and the denominator by the square root which appears in the denominator.

•Example 5:

Rationalize the denominators:

     a. $\dfrac{1}{3\sqrt{2}}$ b. $\dfrac{2}{\sqrt{5}}$ c. $\dfrac{\sqrt{7}}{\sqrt{x}}$

Solution: Our goal is to eliminate the radical in the denominators. In problem a., the denominator contains $\sqrt{2}$. Therefore, multiply the fraction by $\dfrac{\sqrt{2}}{\sqrt{2}}$ to get:

a. $\dfrac{1}{3\sqrt{2}}$ $= \dfrac{1}{3\sqrt{2}} \cdot \dfrac{\sqrt{2}}{\sqrt{2}} = \dfrac{\sqrt{2}}{3 \cdot \sqrt{4}}$ $= \dfrac{\sqrt{2}}{3 \cdot 2} = \dfrac{\sqrt{2}}{6}$

b. $\dfrac{2}{\sqrt{5}}$ $= \dfrac{2}{\sqrt{5}} \cdot \dfrac{\sqrt{5}}{\sqrt{5}} = \dfrac{2\sqrt{5}}{5}$ Since $\sqrt{5} \cdot \sqrt{5} = \sqrt{25} = 5$.

c. $\dfrac{\sqrt{7}}{\sqrt{x}}$ $= \dfrac{\sqrt{7}}{\sqrt{x}} \cdot \dfrac{\sqrt{x}}{\sqrt{x}} = \dfrac{\sqrt{7x}}{x}$ Since $\sqrt{x} \cdot \sqrt{x} = x$.

In the answer to Problem 5b. above, namely $\dfrac{2\sqrt{5}}{5}$, it is illegal to cancel the 5s, since the 5 in the numerator is under the radical and the 5 in the denominator is not.

Often, a problem needs to be simplified (Examples 2, 3, 4) and then rationalized (Example 5). The next example shows how to combine these concepts.

•Example 6:

Simplify $\dfrac{\sqrt{50}}{\sqrt{4}}$.

Solution: First use the Quotient Rule to simplify.

$$\dfrac{\sqrt{50}}{\sqrt{4}} = \sqrt{\dfrac{50}{4}} = \sqrt{\dfrac{25}{2}} = \dfrac{5}{\sqrt{2}}$$

Then rationalize the denominator.

$$\dfrac{5}{\sqrt{2}} = \dfrac{5}{\sqrt{2}} \cdot \dfrac{\sqrt{2}}{\sqrt{2}} = \dfrac{5\sqrt{2}}{2}$$

•Example 7:

Simplify: a. $\dfrac{6}{\sqrt{2}}$ b. $\dfrac{\sqrt{5xyz}}{\sqrt{x^2yz}}$

a. $\dfrac{6}{\sqrt{2}} = \dfrac{6}{\sqrt{2}} \cdot \dfrac{\sqrt{2}}{\sqrt{2}} = \dfrac{6\sqrt{2}}{2} = \dfrac{\overset{3}{\cancel{6}}\sqrt{2}}{\cancel{2}} = 3\sqrt{2}$

b. $\sqrt{\dfrac{5xyz}{x^2yz}} = \sqrt{\dfrac{5x}{x^2}} = \sqrt{\dfrac{5}{x}} = \dfrac{\sqrt{5}}{\sqrt{x}} \cdot \dfrac{\sqrt{x}}{\sqrt{x}} = \dfrac{\sqrt{5x}}{x}$

In Part 7a. above, it is legal to divide the 6 in the numerator by the 2 in the denominator, since both numbers are outside of the radical sign.

Radical Expressions, Exercise 7.

1. Verify that $\sqrt{\dfrac{36}{4}} = \dfrac{\sqrt{36}}{\sqrt{4}}$ by working both sides separately as in Example 1.

2. Use the Quotient Rule to simplify each:

a. $\dfrac{\sqrt{45}}{\sqrt{5}}$ b. $\dfrac{\sqrt{35}}{\sqrt{7}}$ c. $\dfrac{\sqrt{6x}}{\sqrt{24}}$ d. $\dfrac{\sqrt{45x^2}}{\sqrt{9}}$

e. $\sqrt{\dfrac{14x}{18x}}$ f. $\sqrt{\dfrac{75xy^7}{12xy}}$ g. $\sqrt{\dfrac{2x}{9x}}$

3. Rationalize the denominator, as in Example 5.

a. $\dfrac{4}{\sqrt{3}}$ b. $\dfrac{\sqrt{2}}{\sqrt{5}}$ c. $\dfrac{\sqrt{3}}{\sqrt{x}}$ d. $\dfrac{5}{7\sqrt{2}}$

e. $\dfrac{\sqrt{2}}{5\sqrt{3}}$ f. $\dfrac{2}{\sqrt{2}}$ g. $\dfrac{x}{\sqrt{x}}$

4. Simplify, then rationalize denominators. (As in Examples 6 and 7)

a. $\dfrac{\sqrt{25}}{\sqrt{2}}$ b. $\sqrt{\dfrac{18}{10}}$ c. $\dfrac{\sqrt{14}}{\sqrt{21}}$ d. $\sqrt{\dfrac{5x}{15}}$

e. $\dfrac{\sqrt{x}}{\sqrt{y}}$ f. $\dfrac{\sqrt{a^2}}{\sqrt{2a}}$ g. $\sqrt{\dfrac{4x \cdot y \cdot z}{2x^2yz}}$ h. $\sqrt{\dfrac{7}{28x}}$

Radical Equations

An equation that contains a variable radicand is called a **radical equation**. For example, $\sqrt{x} = 3$ and $y + 1 = \sqrt{y + 4}$ are radical equations.

Solving a radical equation means finding values that make the equation true.

Radical equations are solved by first isolating the radical and then squaring both sides.

•Example 1:

> Solve $\sqrt{x} = 3$.
>
> Solution: Our goal is to get rid of the radical sign ($\sqrt{}$). We can eliminate the $\sqrt{}$ by squaring both sides of the equal sign:
>
> $$(\sqrt{x})^2 = 3^2$$
> $$x = 9 \qquad \text{Finished.}$$

Squaring both sides causes the $\sqrt{}$ to go away,

$$\begin{aligned} \text{because } (\sqrt{x})^2 \quad &= \sqrt{x} \cdot \sqrt{x} &&\text{2 factors of } \sqrt{x} \\ &= \sqrt{x \cdot x} &&\text{product rule.} \\ &= x \end{aligned}$$

The solutions to radical equations must always be checked, because the act of squaring both sides of an equation sometimes gives us incorrect results.

The solution to Example 1 checks, because replacing x with 9 in the original equation gives: $\sqrt{x} = 3$, $\sqrt{9} = 3$ which is true.

•Example 2:

> Solve $7\sqrt{x + 5} + 10 = 66$.
>
> Solution: Before squaring both sides, we must isolate the radical; that is, get the $\sqrt{x + 5}$ alone.
>
> $$\begin{aligned} 7\sqrt{x + 5} + 10 &= 66 \\ 7\sqrt{x + 5} &= 56 &&\text{Subtract 10 from both sides.} \\ \sqrt{x + 5} &= 8 &&\text{Divide both sides by 7.} \end{aligned}$$
>
> Now, square both sides to get rid of the radical sign.
>
> $$\begin{aligned} (\sqrt{x + 5})^2 &= 8^2 \\ x + 5 &= 64 \\ x &= 59 \qquad \text{Finished.} \end{aligned}$$

71

Check x = 59 by replacing x with 59 in the original equation:

$$7\sqrt{59 + 5} + 10 = 66$$
$$7\sqrt{64} + 10 = 66$$
$$7 \cdot 8 + 10 = 66$$
$$56 + 10 = 66$$
$$66 = 66 \text{ True.}$$

•Example 3:

Solve $\sqrt{2 - x} - 4 = x$.

Solution: First, isolate $\sqrt{2 - x}$ by adding 4 to both sides:

$$\sqrt{2 - x} = x + 4$$

Now, square both sides to remove the radical sign:

$(\sqrt{2 - x})^2 = (x + 4)^2$

$2 - x = (x + 4)(x + 4)$ 2 factors of (x +4)

$2 - x = x^2 + 8x + 16$ By FOIL

$0 = x^2 + 9x + 14$ Add x, subtract 2.

$0 = (x + 7)(x + 2)$ Factor

$x + 7 = 0$ or $x + 2 = 0$ Zero Product Rule

$x = -7$ or $x = -2$

<u>Check x = -7</u>

$\sqrt{2 - (-7)} - 4 = -7$

$\sqrt{9} - 4 = -7$

$3 - 4 = -7$

$-1 = -7$ False.

x = -7 is not a solution.

<u>Check x = -2</u>

$\sqrt{2 - (-2)} - 4 = -2$

$\sqrt{4} - 4 = -2$

$2 - 4 = -2$

$-2 = -2$ True.

Hence x = -2 is the solution.

The above example suggests these steps:

Step 1: Isolate the radical.
Step 2: Square both sides. If this results in an x2 term, then use the Zero Product Rule as in Example 3.
Step 3: Check the solutions.

Radical Expressions, Exercise 8. Solve these equations.

1. $\sqrt{x} = 10$

2. $\sqrt{x} - 1 = 3$

3. $4\sqrt{x} - 7 = 13$

4. $\sqrt{3x} = 9$

5. $\sqrt{8x} + 3 = 7$

6. $\sqrt{7x + 2} = x + 2$ Hint: Example 3.

7. $\sqrt{x - 1} = x - 7$

8. $\sqrt{2x} = \sqrt{x + 5}$ Hint Square both sides

9. $3\sqrt{x} = \sqrt{4x + 5}$ Hint: $(3\sqrt{x})^2 = 9x$.

10. $\sqrt{x} = -2$ Hint: Check your answer.

Quadratic Equations

Examples of quadratic equations are shown below. Each is of degree 2. An equation is <u>not</u> quadratic if its degree is different than 2.

<u>Quadratic Equations</u>

$x^2 = 9$
$5x^2 + 10 = 3x$
$7(x + 1)^2 = 5$
$3 + (x - 2)^2 = 5$

<u>Not Quadratic</u>

$x^3 = 9$ (degree 3)
$4x^5 - x^2 = 4$ (degree 5)
$(x + 1)^4 = 10$ (degree 4)
$2x + 1 = 5$ (degree 1, linear)

To solve a quadratic equation means to find all the values that make the equation true. In this chapter, you will learn <u>three methods</u> for solving quadratic equations. The first method is called the square root method.

The Square Root Method (Method 1)

This procedure is called the square root method, because it involves taking the square root on both sides of the equal sign.

•Example 1:

> Solve $x^2 = 9$.
>
> Solution: Our goal is to get x completely alone by eliminating the exponent 2. We eliminate the 2 by taking the square root of both sides and writing the symbol ± in front of the non-variable side:
>
> $$x^2 = 9$$
> $$\sqrt{x^2} = \pm \sqrt{9}$$
> $$x = \pm 3 \qquad \text{Since } \sqrt{x^2} = \sqrt{x \cdot x} = x \text{ and } \sqrt{9} = 3$$
>
> The symbol ± means **plus or minus** and the final result, $x = \pm 3$, means the solutions are $x = +3$ or $x = -3$.
>
> The checks below show that this procedure does indeed give the correct solutions:
>
> Check $x = +3$
>
> $x^2 = 9$
> $3^2 = 9$
> $9 = 9$ True.
>
> Check $x = -3$
>
> $x^2 = 9$
> $(-3)^2 = 9$
> $9 = 9$ True.

In the above example, it was essential to write the symbol ± in front of $\sqrt{9}$. Otherwise, we would not have discovered that -3 is also a solution.

•Example 2:

Solve $2x^2 - 9 = 15$.

Solution: Before we take the square root on both sides, we need to isolate x^2.

$$2x^2 - 9 = 15$$
$$2x^2 = 24 \qquad \text{Adding 9 to both sides.}$$
$$x^2 = 12 \qquad \text{Dividing both sides by 2.}$$

Now take the square root of both sides and write the symbol ±.
$$x^2 = 12$$
$$\sqrt{x^2} = \pm \sqrt{12}$$
$$x = \pm 2\sqrt{3} \qquad \text{Since } \sqrt{12} = \sqrt{4 \cdot 3} = 2\sqrt{3}$$

Therefore, $x = +2\sqrt{3}$ and $x = -2\sqrt{3}$ are the solutions to $2x^2 - 9 = 15$.

•Example 3:

Solve $(3x + 2)^2 + 4 = 29$.

Solution: First get $(3x + 2)^2$ alone by subtracting 4 from both sides.
$$(3x + 2)^2 + 4 = 29$$
$$(3x + 2)^2 = 25$$

Now, take the square root of both sides.
$$\sqrt{(3x + 2)^2} = \pm \sqrt{25}$$
$$3x + 2 = \pm 5 \qquad \text{which means that:}$$

Either $3x + 2 = +5$ or $3x + 2 = -5$.

Solve each of these equations separately to get:

$$3x + 2 = +5 \qquad\qquad\qquad\qquad 3x + 2 = -5$$
$$3x = 3 \qquad \text{Subtract 2.} \qquad 3x = -7$$
$$x = 1 \qquad \text{Divide by 3.} \qquad x = -\frac{7}{3}$$

•Example 4:

Solve $2(5x - 1)^2 = 36$.

Solution: $(5x - 1)^2 = 18$ Divide both sides by 2.

$$\sqrt{(5x - 1)^2} = \pm \sqrt{18}$$
$$5x - 1 = \pm 3\sqrt{2}$$
$$5x = 1 \pm 3\sqrt{2}$$ Add one to both sides.
$$x = \frac{1 \pm 3\sqrt{2}}{5}$$ Divide both sides by 5.

In the above example, $x = \dfrac{1 \pm 3\sqrt{2}}{5}$ means that the solutions are $x = \dfrac{1 + 3\sqrt{2}}{5}$ and $x = \dfrac{1 - 3\sqrt{2}}{5}$. These expressions cannot be simplified since 1 and $3\sqrt{2}$ are unlike terms.

Quadratic Equations, Exercise 1. Solve.

1. $x^2 = 64$
2. $x^2 = 12$
3. $x^2 - 50 = 0$
4. $2x^2 = 32$

5. $2x^2 + 3 = 11$
6. $5x^2 - 7 = -2$
7. $(x + 3)^2 = 16$
8. $2(x - 1)^2 - 10 = 8$

9. $(3x - 2)^2 - 4 = 0$
10. $(2x + 1)^2 = 12$
11. $(3x - 5)^2 = 18$
12. $x^2 = -4$

13. In Example 4, use a calculator to calculate $\dfrac{1 + 3\sqrt{2}}{5}$ and $\dfrac{1 - 3\sqrt{2}}{5}$.

14. In Example 2, check that $2\sqrt{3}$ is a solution to $2x^2 - 9 = 15$ by replacing x with $2\sqrt{3}$. Hint: $(2\sqrt{3})^2 = (2\sqrt{3}) \cdot (2\sqrt{3}) = 2 \cdot 2 \cdot \sqrt{3} \cdot \sqrt{3}$.

The Zero Product Method (Method 2)

You are already familar with this method. We have used it previously.
Step 1: Get zero on one side of the equal sign.
Step 2: Factor.
Step 3: Set each factor equal to zero and solve.

•Example 1:

Solve $3x^2 = 2x$.

Step 1:	$3x^2 - 2x = 0$	Subtract 2x from both sides.
Step 2:	$x(3x - 2) = 0$	
Step 3:	either $x = 0$	or $3x - 2 = 0$
	$x = 0$	or $3x = 2$
	$x = 0$	or $x = \frac{2}{3}$

76

•Example 2:

Solve $3 - 2x^2 = 5x$.

Step 1: $0 = 2x^2 + 5x - 3$ Adding $2x^2$, subtracting 3.
Step 2: $0 = (2x - 1)(x + 3)$
Step 3:   either $2x - 1 = 0$ or $x + 3 = 0$
                      $2x = 1$ or $x = -3$
                      $x = \frac{1}{2}$ or $x = -3$

Quadratic Equations, Exercise 2. Solve.

1. $x^2 + 5x = 0$
2. $x^2 - x = 0$
3. $7x^2 = 5x$
4. $x^2 - 25 = 0$
   Hint: $(x + 5)(x - 5) = 0$

5. $x^2 - 4 = 0$
6. $4x^2 - 49 = 0$
7. $2x^2 = 32$
   Hint: Divide both sides by 2 first.

8. $x^2 - 4x - 21 = 0$
9. $3x^2 + 6x = -3$
   Hint: divide by 3 first.
10. $x + 6 = 2x^2$

The Quadratic Formula Method (Method 3)

To solve a quadratic equation by this method, it is first necessary to rewrite the equation in **standard form**.

The **standard form** of a quadratic equation is **$ax^2 + bx + c = 0$**; where **a** represents the coefficient of the **x^2** term, **b** represents the coefficient of the **x** term, and **c** represents the coefficient of the **non-variable** term.

•Example 1:

Identify a, b, and c in $5x^2 + 3x - 2 = 0$.

Solution: $5x^2 + 3x - 2 = 0$ is in the standard form $ax^2 + bx + c = 0$. Therefore, a = 5, b = 3, and c = -2.

•More examples:

Standard Form	Coefficients		
$7x^2 + 10x + 9 = 0$	a = 7	b = 10	c = 9
$-3x^2 + x + 5 = 0$	a = -3	b = 1	c = 5
$x^2 - x + 0 = 0$	a = 1	b = -1	c = 0
$x^2 - 2x = 0$	a = 1	b = -2	c = 0
$4x^2 + 0x + 9 = 0$	a = 4	b = 0	c = 9
$-x^2 + 5 = 0$	a = -1	b = 0	c = 5

When an equation is in standard form, the solutions can be found by using the **Quadratic Formula**

$$x = \frac{-b \pm \sqrt{b^2 - 4ac}}{2a}$$

•Example 2:

Solve $x^2 - 4x + 3 = 0$.

Solution: This equation is in the standard form with **a** = 1, **b** = -4, and **c** = +3. Thus in the quadtaric formula, replace **a** with 1, **b** with -4, and **c** with 3.

$$x = \frac{-b \pm \sqrt{b^2 - 4ac}}{2a}$$

$$x = \frac{-(-4) \pm \sqrt{(-4)^2 - 4(1)(3)}}{2(1)} \qquad \text{Since a = 1, b = -4, c = 3.}$$

$$x = \frac{4 \pm \sqrt{16 - 12}}{2} = \frac{4 \pm \sqrt{4}}{2} = \frac{4 \pm 2}{2}$$

Therefore either $x = \dfrac{4 + 2}{2}$ or $x = \dfrac{4 - 2}{2}$

$$x = \frac{6}{2} \qquad\qquad x = \frac{2}{2}$$

$$x = 3 \qquad \text{or} \qquad x = 1$$

To solve Using the Quadratic Formula:
Step 1: Rewrite the equation in standard form.
Step 2: Identify a, b, and c.
Step 3: Write down the quadratic formula. Then replace a, b, c.

•Example 3:

Solve $2x = 15 - x^2$.

Step 1: Get zero on one side by subtracting 15 and adding x^2.

$$2x = 15 - x^2$$
$$2x - 15 + x^2 = 0$$
$$x^2 + 2x - 15 = 0 \quad \text{Rearrange in decending order.}$$

Step 2:   a = 1,   b = 2,   c = -15.

Step 3: $x = \dfrac{-b \pm \sqrt{b^2 - 4ac}}{2a}$

$x = \dfrac{-2 \pm \sqrt{2^2 - 4(1)(-15)}}{2(1)}$

$x = \dfrac{-2 \pm \sqrt{4 + 60}}{2}$ Since $-4(1)(-15) = +60$.

$x = \dfrac{-2 \pm \sqrt{64}}{2} = \dfrac{-2 \pm 8}{2}$

Thus either, $x = \dfrac{-2 + 8}{2}$ or $x = \dfrac{-2 - 8}{2}$

$x = \dfrac{6}{2}$ $x = \dfrac{-10}{2}$

$x = 3$ or $x = -5$

•Example 4:

Solve $3x^2 = 1$.

Step 1: $3x^2 - 1 = 0$ Standard form.
Step 2: $a = 3$, $b = 0$, $c = -1$.
     $b = 0$ because there is no x term.

Step 3: $x = \dfrac{-b \pm \sqrt{b^2 - 4ac}}{2a}$

$x = \dfrac{-0 \pm \sqrt{0^2 - 4(3)(-1)}}{2(3)} = \dfrac{0 \pm \sqrt{0 + 12}}{6}$
     Since $-4(3)(-1) = +12$

$x = \dfrac{0 \pm \sqrt{12}}{6} = \dfrac{0 \pm 2\sqrt{3}}{6}$

Hence, either $x = \dfrac{0 + 2\sqrt{3}}{6}$ or $x = \dfrac{0 - 2\sqrt{3}}{6}$

$x = \dfrac{2\sqrt{3}}{6}$ or $x = \dfrac{-2\sqrt{3}}{6}$

$x = \dfrac{\sqrt{3}}{3}$ or $x = -\dfrac{\sqrt{3}}{3}$

     By dividing both the
     2 and the 6 by 2.

In Example 4, the answers could be displayed as $\pm \dfrac{\sqrt{3}}{3}$.

All examples thus far had two solutions. Example 5, has one solution. Example 6 has none.

•Example 5:

> Solve $x^2 + 6x + 9 = 0$.
>
> Solution: $a = 1$, $b = 6$, $c = 9$
>
> $$x = \frac{-b \pm \sqrt{b^2 - 4ac}}{2a}$$
>
> $$x = \frac{-6 \pm \sqrt{6^2 - 4(1)(9)}}{2(1)} = \frac{-6 \pm \sqrt{36 - 36}}{2} = \frac{-6 \pm \sqrt{0}}{2}$$
>
> $$= \frac{-6 \pm 0}{2} = \frac{-6}{2} = -3$$
>
> only one solution, $x = -3$.

The above example has only one solution, since adding or subtracting 0 has no affect on the numerator. That is, -6 + 0 = -6 and -6 - 0 = -6.

•Example 6:

> Solve $x^2 + x + 5 = 0$.
>
> Solution: $a = 1$, $b = 1$, $c = 5$.
>
> $$x = \frac{-b \pm \sqrt{b^2 - 4ac}}{2a}$$
>
> $$x = \frac{-1 \pm \sqrt{1^2 - 4(1)(5)}}{2(1)} = \frac{-1 \pm \sqrt{1 - 20}}{2} = \frac{-1 \pm \sqrt{-19}}{2}$$
>
> But $\sqrt{-19}$ has no meaning, since the radicand is negative. Therefore $x^2 + x + 5 = 0$ has no solution.

The **discriminant** is the expression inside the radical sign of the quadratic formula, namely $b^2 - 4ac$. You can tell how many solutions an equation will have by calculating the discriminant.

If $b^2 - 4ac$ is positive, then the equation will have 2 solutions.
If $b^2 - 4ac$ is zero, then the equation will have 1 solution.
If $b^2 - 4ac$ is negative, then the equation will have no solution.

•Example 7:

> Without solving, tell how many solutions each equation will have.
>
> a. $x^2 + 7x + 1 = 0$ b. $x^2 + 2x + 1 = 0$ c. $3x^2 + 2x + 4 = 0$
>
> Solution: a. In $x^2 + 7x + 1 = 0$, $a = 1$, $b = 7$, $c = 1$. The discriminant = $b^2 - 4ac = 7^2 - 4(1)(1) = 49 - 4 = 45$. Since the discriminant is positive, the equation will have two solutions.
>
> b. $b^2 - 4ac = 2^2 - 4(1)(1) = 4 - 4 = 0$. Since the discriminant is 0, the equation, $x^2 + 2x + 1 = 0$, will have one solution.
>
> c. $b^2 - 4ac = 2^2 - 4(3)(4) = 4 - 48 = -44$. A negative discriminant indicates that $3x^2 + 2x + 4 = 0$ has no solution.

Quadratic Equations, Exercise 3. Use the quadratic formula to solve.

1. $x^2 + 2x - 3 = 0$
2. $2x^2 - 3x + 1 = 0$
3. $x^2 + 6x + 9 = 0$
4. $x^2 + 5 = 0$ (b = 0)
5. $x^2 - x = 0$ Hint: b = -1 and c = 0.
6. $4x^2 - 4x = -1$
7. $5x^2 = 4x - 1$
8. $x^2 + 2 = 7x$
9. $3x^2 = 15$ Divide by 3.
10. $2x^2 = -4$
11. $x^2 + 4x - 1 = 0$

For problems 13 through 18, use the discriminant $b^2 - 4ac$ to determine the number of solutions, as in Example 7.

12. $4x^2 + 4x + 1 = 0$
13. $9x^2 - x + 2 = 0$
14. $x^2 + 3x + 1 = 0$
15. $x^2 - x + 5 = 0$
16. $x^2 + 2 = 5x$
17. $x^2 - 6x = -9$

For problems 19 through 21, you will need to refer to the examples and identify the radicands.

18. In Example 2, the discriminant = ___, indicating ___ solutions.
19. In Example 5, the discriminant = ___, indicating ___ solutions.
20. In Example 6, the discriminant = ___, indicating ___ solutions.
21. In Example 7, part b., namely $x^2 + 2x + 1 = 0$, it was found that the discriminant was 0, indicating one solution. Use the quadratic formula to find this one solution.

Summary of Solving Quadratic Equations

The three methods of solving quadratic equations are outlined below:

Case 1: The square root method can be used to solve problems like $x^2 = 25$, $(x - 3)^2 = 4$ or $(2x + 1)^2 = 10$. That is, if you can rewrite the equation with a squared expression on one side and a number on the other side, then use the **square root method**.

Case 2: If you can rewrite the equation in standard form, $ax^2 + bx + c = 0$, then try the **zero product method**. If the expression does not factor, then use the **quadratic formula method**.

For further examples, refer to previous sections.

Quadratic Equations, Exercise 4. Solve each using the most convenient method.

1. $3(x - 2)^2 = 12$ 2. $x^2 - 3x + 2 = 0$ 3. $x^2 + 5x + 2 = 0$

4. Solve $x^2 = 25$ by:
   a. The square root method;
   b. The zero product method;
   c. The quadratic formula.

5. Solve $6x^2 + x - 2 = 0$ by: (1) the Zero Product Rule and (2) the Quadratic Formula.

Graphing Quadratics (Parabolas)

In this section, we will graph quadratic equations by the point plotting method. This process is similar to the point plotting method in Chapter 1.
Step 1: Make an x, y-table.
Step 2: Choose 5 or 6 values at random for x. In the equation, replace x by these values and solve for y.
Step 3: Record the results on the table and plot the points on a graph.

•Example 1:

Graph $y = x^2 - 4x + 3$.

X	Y
0	3
1	0
2	-1
3	0
-1	8
4	4

When x = 0.
$y = x^2 - 4x + 3$
$y = 0^2 - 4(0) + 3$
$y = 3$

When x = 1.
$y = x^2 - 4x + 3$
$y = 1^2 - 4(1) + 3$
$y = 1 - 4 + 3$
$y = 0$

When x = 2
$y = x^2 - 4x + 3$
$y = 2^2 - 4(2) + 3$
$y = 4 - 8 + 3$
$y = -1$

When x = 3.
$y = x^2 - 4x + 3$
$y = 3^2 - 4(3) + 3$
$y = 9 - 12 + 3$
$y = 0$

When x = -1
$y = x^2 - 4x + 3$
$y = (-1)^2 - 4(-1) + 3$
$y = 1 + 4 + 3$
$y = 8$

When x = 4
$y = x^2 - 4x + 3$
$y = 4^2 - 4(4) + 3$
$y = 16 - 16 + 3$
$y = 3$

The ordered pairs in the table are:
(0,3), (1,0), (2,-1), (3,0), (-1,8), and (4,3). Plot these and connect them with a smooth line.

82

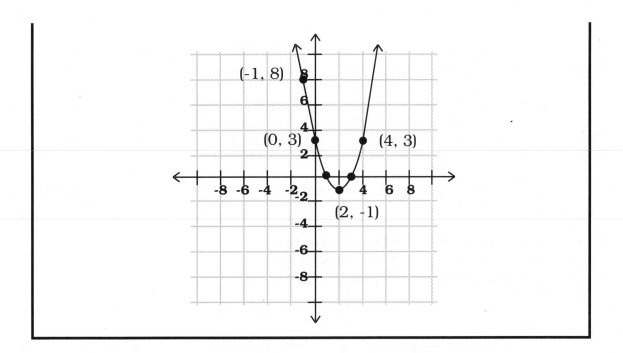

The curved shape in the above example is called a **parabola**.

When quadratic equations of the form $y = ax^2 + bx + c$ are graphed, the result is always a parabola. If the coefficient **a** is positive, then the parabola opens upward. If the coefficient **a** is negative, then the parabola opens downward.

When **a** is positive.

opens upward

When **a** is negative.

opens downward

If a parabola opens upward, its lowest point is called the **vertex**. If a parabola opens downward, its highest point is called the **vertex**. In Example 1, the **vertex** is at point (2,-1)

•Example 2:

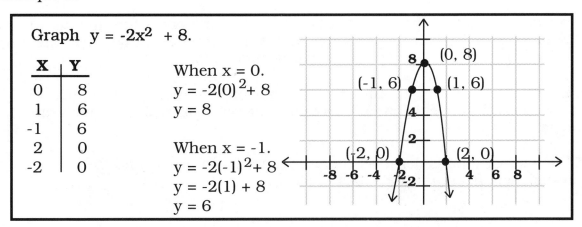

Graph $y = -2x^2 + 8$.

X	Y
0	8
1	6
-1	6
2	0
-2	0

When x = 0.
$y = -2(0)^2 + 8$
$y = 8$

When x = -1.
$y = -2(-1)^2 + 8$
$y = -2(1) + 8$
$y = 6$

In the above example, a = -2. Since **a** is negative the parabola opens downward. The vertex is at the point (0,8).

The **x-intercepts** are the places on the graph where the parabola crosses the x-axis. The **y-coordinate** of an x-intercept is **always zero**. For instance, in Example 2, the x-intercepts are (-2,0) and (2,0) and the y-coordinate of each x-intercept is zero.

The next example shows how the quadratic formula can be used to find the x-intercepts.

To Find the x-Intercepts without Graphing
Step 1: Replace y by 0 (since the y-coordinate = 0).
Step 2: Use the quadratic formula to solve for x.

•Example 3:

Without graphing, find the x-intercepts of $y = -2x^2 + 8$.

Step 1: $0 = -2x^2 + 8$ Replace y by 0.
Step 2: $a = -2, \ b = 0, \ c = 8$

$$x = \frac{-b \pm \sqrt{b^2 - 4ac}}{2a}$$

$$x = \frac{-0 \pm \sqrt{0^2 - 4(-2)(8)}}{2(-2)} = \frac{0 \pm \sqrt{0 + 64}}{-4} = \frac{0 \pm \sqrt{64}}{-4} = \frac{0 \pm 8}{-4}$$

Either $x = \frac{0 + 8}{-4}$ or $x = \frac{0 - 8}{-4}$

$x = -2$ $x = 2$

Therefore the x-intercepts are (-2,0) and (2,0). These are identical to the x-intercepts found by graphing $y = -2x^2 + 8$ in Example 2.

We said before that the discriminant, $b^2 - 4ac$, indicates the number of solutions to the quadratic formula. If the discriminant is positive, there are two solutions and, consequently, two x-intercepts.

If the discriminant is zero, then there is only one solution. Consequently, the graph will have only one x-intercept. The parabola will just touch the x-axis.

If the discriminant is negative, then there are no solutions and the parabola will not cross the x-axis at all.

These relationships are diagrammed in the following chart.

When $y = ax^2 + bx + c$.

Discriminant	Quadratic Formula	Graphs
$b^2 - 4ac$	$x = \dfrac{-b \pm \sqrt{b^2 - 4ac}}{2a}$	

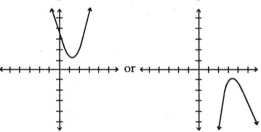

Positive	2 solutions. Graph will cross the x-axis at these 2 values.	
Zero	1 solution. Graph will just touch the x-axis at this one value.	
Negative	No solutions. Graph will not cross the x-axis.	

Quadratic Equations, Exercise 5.

In problems 1, 2, and 3, graph each equation and identify the vertex.

1. $y = x^2$ 2. $y = x^2 + 1$ 3. $y = x^2 - 2x - 3$

4. In Exercise 3, a = 1, b = -2, and c = -3. Calculate the discriminant, $b^2 - 4ac$. Does this discriminant agree with the number of x-intercepts in your graph of Exercise 3? (see chart)

5. In Exercise 3, replace y by 0 and use the quadratic formula to find the x-intercept (see Example 3). Are these x-intercepts identical to those on your graph of Exercise 3?

6. In Exercise 2, $y = x^2 + 1$, a = 1, b = 0, c = 1. Calculate $b^2 - 4ac$. Does this agree with the number of x-intercepts on your graph of Exercise 2?

7. In $y = x^2$, a = 1, b = 0, c = 0. Calculate $b^2 - 4ac$ and compare this result to the number of x-intercepts on your graph of $y = x^2$ (Exercise 1).

8. Graph: (a) $y = -2x^2 + 4$ and (b) $y = -x^2$ (Hint: $-x^2 = -1 \cdot x^2$).

Review

Chapter 4

1. Simplify:

    A. $\sqrt{64}$ B. $-\sqrt{9}$ C. $\sqrt{27}$ D. $\sqrt{72}$ E. $\sqrt{x^6}$

    F. $\sqrt{12y^4}$ G. $\sqrt{x^5}$ H. $\sqrt{3y^7}$ I. $\sqrt{18x^{10}y}$ J. $\sqrt{6^2}$

    K. $\sqrt{x \cdot x}$ L. $\sqrt{500}$ M. $\sqrt{392}$

2. Multiply:

    A. $\sqrt{2} \cdot \sqrt{10}$ B. $\sqrt{7} \cdot \sqrt{7}$ C. $\sqrt{3} \cdot \sqrt{15x}$

    D. $\sqrt{x} \cdot \sqrt{x}$ E. $\sqrt{y} \cdot \sqrt{y^3}$ F. $\sqrt{4xy} \cdot \sqrt{12xy}$

3. Combine like terms.

    A. $5\sqrt{x} + 7\sqrt{x}$ B. $\sqrt{18} + \sqrt{2}$ C. $\sqrt{27x} - \sqrt{48x}$

4. Simplify. Rationalize the denominator when needed.

    A. $\dfrac{\sqrt{4}}{\sqrt{9}}$ B. $\sqrt{\dfrac{27}{48}}$ C. $\sqrt{\dfrac{4x^2}{16}}$ D. $\dfrac{\sqrt{6xy}}{\sqrt{3xy}}$ E. $\dfrac{5}{\sqrt{2}}$

    F. $\sqrt{\dfrac{14}{4}}$ G. $\dfrac{10}{\sqrt{x}}$ H. $\dfrac{3}{\sqrt{3}}$ I. $\dfrac{\sqrt{6x}}{\sqrt{12}}$ J. $\dfrac{5y}{\sqrt{y}}$

5. Solve:

    A. $\sqrt{x} = 5$ B. $2\sqrt{y} - 11 = 1$ C. $\sqrt{3x + 7} + 2 = 3$

Chapter 5

1. Use the square root method to solve:

    A. $x^2 = 8$ B. $7x^2 - 10 = -3$ C. $2(x - 3)^2 - 5 = 13$

2. Use the zero product or quadratic formula to solve:

    A. $x^2 - 8x = 0$ B. $x^2 - 4 = 0$ C. $x^2 - 7x + 12 = 0$

    D. $3x^2 - 5x = 2$ E. $x^2 + 3x + 1 = 0$ F. $x^2 + 7 = 0$

    G. $x^2 + x = 1$ H. $2x^2 - 3x + 1 = 0$ I. $21x^2 + 41x + 10 = 0$

3. Calculate the discriminant, $b^2 - 4ac$, to determine the number of solutions to each.

    A. $0 = x^2 + 7x + 7$ B. $0 = x^2 + x + 9$ C. $0 = x^2 + 10x + 25$

    D. $0 = 25x^2 - 30x + 9$ E. $0 = 4x^2 - 4$ F. $0 = 10x^2 + 5x$

4. Graph:

    A. $y = 2x^2$ B. $y = x^2 - 6x + 9$ C. $y = -x^2$

5. Without graphing, find the x-intercepts of $y = 2x^2 - 9x + 4$.

Cumulative Review

Chapter 1

1. Determine the slope and y-intercept of each:

    A. $y = -3x + 7$ B. $y = \frac{4}{5}x - 6$ C. $y = -x$

    D. $y - 5x = 1$ (solve for y first)   E. $2y = 6x$ F. $\frac{1}{2}y = x + 8$

2. Graph $y = 2x - 4$ by the point plotting method.

3. Graph $y = 2x - 4$ by using zeros.

4. Graph $y = 2x - 4$ by the slope intercept method.

5. Find the slope of a line which passes through the points (-5, -3) and (0, 2).

6. Write the equation of a line which pases through (0,1) with a slope of -3.

Chapter 2

7. Solve $x + y = 0$ by the graphing method.
              $x - y = 0$

8. Solve the system in Exercise 7 by substitution.

9. Solve the system in Exercise 7 by the addition method.

10. Solve: $x + 2y = -5$
                $2x - y = 5$

       A. Using graphing   B. Using substitution   C. Using addition

11. Solve: $x + y = 2$ By any method.
                $2x + 2y = 3$

12. Solve: $2x - 6y = 2$ By any method.
                $3x - 9y = 3$

Chapter 3

13. Reduce:

    A. $\dfrac{x}{xy}$
        B. $\dfrac{2x + 4}{4}$
        C. $\dfrac{x - 3}{5x - 15}$

    D. $\dfrac{x^2 + 9x}{x}$
        E. $\dfrac{x - 2}{x^2 - 2x}$
        F. $\dfrac{x + 1}{x^2 + 7x + 6}$

14. Multiply and divide:

    A. $\dfrac{x}{x + 7} \cdot \dfrac{x + 7}{1}$
        B. $\dfrac{x}{2x + 10} \cdot \dfrac{2}{x}$

    C. $\dfrac{x^2 + 9x + 14}{(x + 1)} \div \dfrac{(x + 7)}{(x + 1)}$
    D. $\dfrac{y^2 - y}{1} \div \dfrac{y^2 - 1}{3}$

15. Add or subtract:

    A. $\dfrac{1}{x} + \dfrac{1}{2}$
        B. $\dfrac{1}{x - 5} - \dfrac{5}{x(x - 5)}$
        C. $\dfrac{x}{x^2 + 2x + 1} + \dfrac{1}{x + 1}$

16. Simplify:
    $\dfrac{\frac{1}{x} + \frac{1}{2}}{\frac{1}{2x}}$.

17. Solve:
    $\dfrac{1}{x} + \dfrac{2}{3} = \dfrac{1}{3x}$.

Chapter 4

18. Simplify: A. $\sqrt{100}$ B. $\sqrt{0}$ C. $\sqrt{8}$ D. $\sqrt{20}$

19. Multiply: A. $\sqrt{30} \cdot \sqrt{18}$ B. $\sqrt{p} \cdot \sqrt{p}$ C. $\sqrt{21x} \cdot \sqrt{14x}$

20. Add or subtract: A. $\sqrt{18} - \sqrt{8}$ B. $\sqrt{25x} + \sqrt{36x}$

21. Simplify: A. $\sqrt{\dfrac{20}{5}}$ B. $\dfrac{\sqrt{5xy}}{\sqrt{xy}}$ C. $\dfrac{7}{\sqrt{7}}$ D. $\dfrac{\sqrt{15}}{\sqrt{3x}}$

22. Solve: $3\sqrt{6x + 1} = 15$.

Chapter 5

23. Use the square root method to solve:

    A. $x^2 = 100$ B. $(2x - 1)^2 = 9$ C. $(x + 2)^2 - 10 = 0$

24. Use the zero product method to solve:

    A. $3x^2 - 21x = 0$ B. $6x^2 + x - 2 = 0$ C. $x^2 - 49 = 0$

25. Use the quadratic formula to solve:

    A. $x^2 - 5x + 3 = 0$ B. $3x^2 - 1 = 0$ C. $x^2 + 7 = 0$

26. Use the most appropriate method to solve:

    A. $x^2 = 50$ B. $x^2 + 11x + 18 = 0$ C. $x^2 + 3x + 1 = 0$

    D. $4x^2 - 9 = 0$ E. $x^2 + x = -1$ F. $2x^2 = 32$

27. Graph:

    A. $y = x^2 - 3$ B. $y = x^2 + 1$ C. $y = x^2 - 2x + 1$

28. Calculate the discriminant of:

    A. $0 = x^2 - 3$ B. $0 = x^2 + 1$ C. $0 = x^2 - 2x + 1$

    Compare these results to your graphs in Exercise 27.

Answers

Graphing, Exercise 1, page 3.

A. (4, 1)   B. (1, 4)   C. (-3,6)   D. (-5, -2)   E. (2, -2)
F. (3, -5)   G. (-3, 0)   H. (0, 0)   I. (0, 6)

Graphing, Exercise 3, page 4.

1a. yes	b. yes	c. no
d. yes	e. no	f. yes
2a. yes	b. yes	c. yes
d. no	e. yes	f. no
3a. yes	b. no	c. yes
d. no	e. no	f. yes
4a. yes	b. no	c. yes
d. yes	e. yes	

Graphing, Exercise 2, page 3.

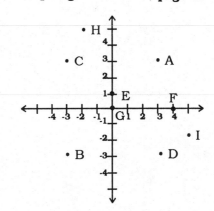

Graphing, Exercise 4, page 8.

1.
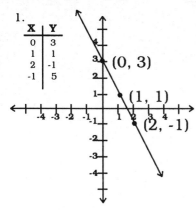

2.
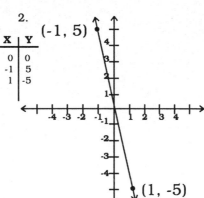

3.
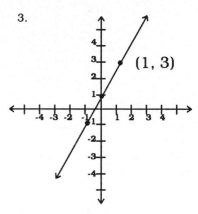

4.
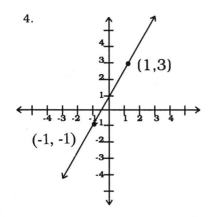

5.
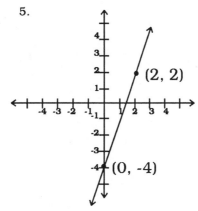

6.
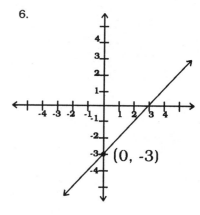

7.
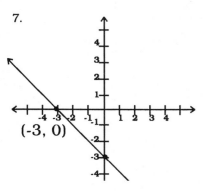

8.
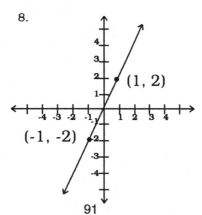

9.
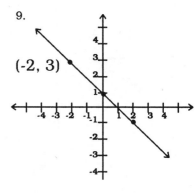

91

10.

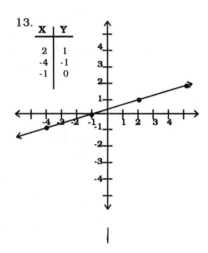

X	Y
0	1
-2	2
2	0

11.

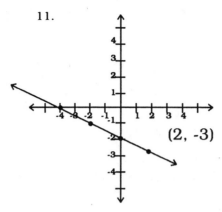

(2, -3)

12.

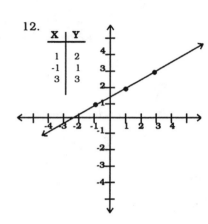

X	Y
1	2
-1	1
3	3

13.

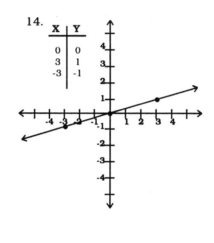

X	Y
2	1
-4	-1
-1	0

14.

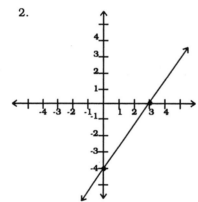

X	Y
0	0
3	1
-3	-1

15.

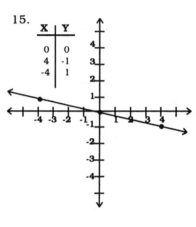

X	Y
0	0
4	-1
-4	1

Graphing, Exercise 5, page 9.

1.

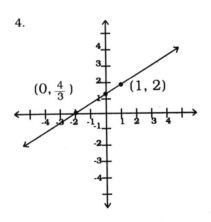

(0, 3)

(5, 0)

2.

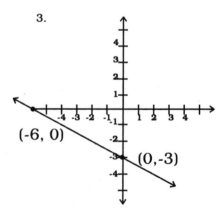

3.

(-6, 0)

(0, -3)

4.

(0, $\frac{4}{3}$)

(1, 2)

5. In Problem 1, x-intercept = 5, y intercept = 3.
   In Problem 2, x-intercept = 3, y-intercept = -4.
   In Problem 3, x-intercept = -6, y-intercept = -3.
   In Problem 4, x-intercept = -2, y-intercept = $\frac{4}{3}$ = $1\frac{1}{3}$.

Graphing Exercise 6, page 11,

1.
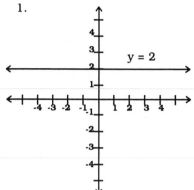
y = 2

2.
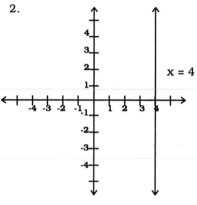
x = 4

3.
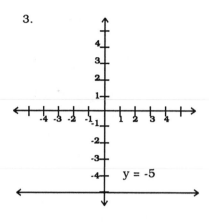
y = -5

4. The graph of x = 0 is a line which is precisely on top of the y-axis.

5.
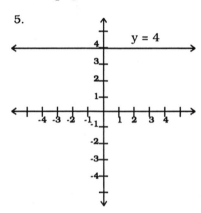
y = 4

6.
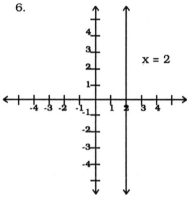
x = 2

Graphing Exercise 7, page 15.

1. $\frac{2}{6} = \frac{1}{3}$ 2. $\frac{-3}{6} = -\frac{1}{2}$ 3. $-\frac{3}{3} = -1$ 4. $\frac{4}{1} = 4$ 5. $\frac{0}{3} = 0$

Graphing Exercise 8, page 15.

1. 1   2. $\frac{7}{3}$ 3. $\frac{3}{5}$ 4. 0   5. $\frac{6}{4} = \frac{3}{2}$

Graphing, Exercise 9, 19.

1. y-intercept = -1
   slope = $\frac{3}{2}$
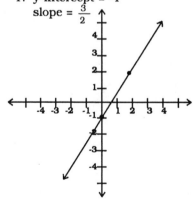

2. y-intercept = 2
   slope = $\frac{-1}{3}$
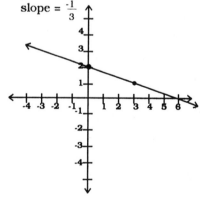

3. y-intercept = -4
   slope = 2
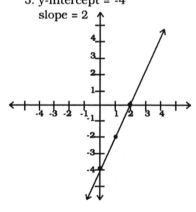

93

4. y-intercept = 4
slope = -2

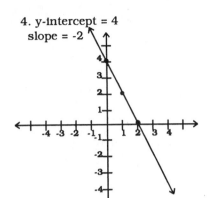

5. y-intercept = 0
slope = $\frac{1}{2}$

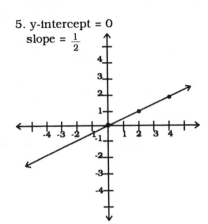

6. y-intercept = 2
slope = 1

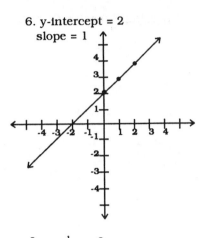

7. y-intercept = 2
slope = -1

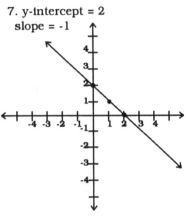

8. y-intercept = -3
slope = $\frac{4}{5}$

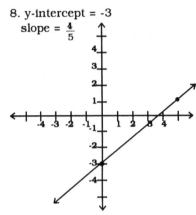

9. $y = \frac{1}{3} x + 2$

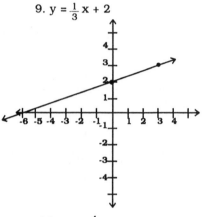

10. $y = 2x + 1$

y-intercept = 1
slope = 2

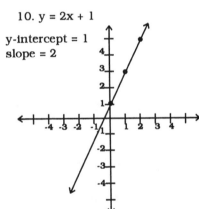

11. $y = -\frac{1}{3} x$

y-intercept = 0
slope = $-\frac{1}{3}$

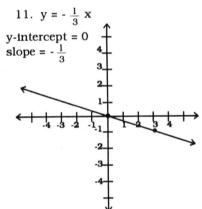

12. $y = -\frac{4}{3} x$

y-intercept = 0
slope = $-\frac{4}{3}$

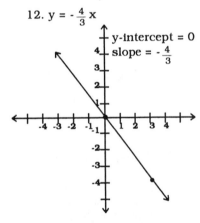

13. $5y = -2x + 15$

$y = -\frac{2}{5} x + 3$

slope = $-\frac{2}{5}$

y-intercept = 3

14. $y = -2x + 5$

y-intercept = 5

slope = $-\frac{2}{1}$

15. $y = -2x - 5$

y intercept = -5

slope = -2

16. $y = \frac{1}{3} x + \frac{2}{3}$

y-intercept = $\frac{2}{3}$

slope = $\frac{1}{3}$

Graphing, Exercise 10, page 20.

1. $y = x + 1$ 2. $y = 2x - 17$ 3. $y = -4x - 7$ 4. $y = -x + 6$ 5. $y = x$
6a. Slope = 8   b. $2 = 8 \cdot 1 + b$ c. $b = -6$

Systems, Exercise 1, page 22.

1a. Yes   b. No   c. No   2a. No   b. Yes   c. No   3a. No   b. No   c. Yes
4a. Yes   b. Yes   c. Yes   5a. No   b. Yes   c. No

Systems, Exercise 2, page 25.

1. Inconsistent   2. Consistent   3. Dependent   4. Consistent   5. Dependent   6. Inconsistent

7. Consistent: solution (-1, 2).

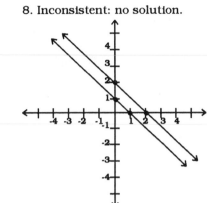

8. Inconsistent: no solution.

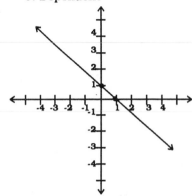

9. Dependent

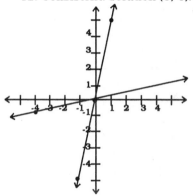

10. Consistent, solution (1, 4).

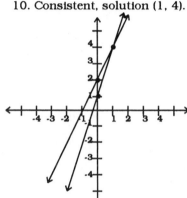

11. Inconsistent.

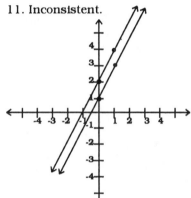

12. Consistent: solution (0, 0).

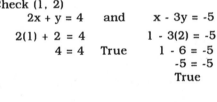

13. Dependent.

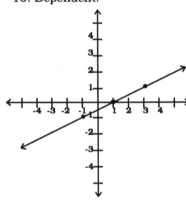

14. Consistent: solution (-1, 1).

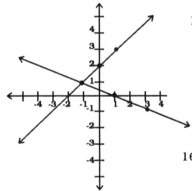

15. Check (1, 2)

$2x + y = 4$ and $x - 3y = -5$

$2(1) + 2 = 4$ $1 - 3(2) = -5$

$4 = 4$ True $1 - 6 = -5$

$-5 = -5$

True

16. $x + 2y = 2$ First equation.

$3x + 3(2y) = 3 \cdot 2$ Multiply 3 to both sides.

$3x + 6y = 6$ Same as second equation.

Hence, lines are identical when graphed.

Systems, Exercise 3, page 30.

1. (-1, 1) Hints: Step 1, $y = 4x + 5$. Step 2, $3x + 2(4x + 5) = -1$ 2. (2, -1) Hint: Step 1, $y = 3 - 2x$ or $y = -2x + 3$.   3. (0, 2)   4. (-3, 4) Hint: $(y - 7) + 2y = 5$.   5. (2, -3) Hints: Step 1, $x = y + 5$. Step 2, $2(y + 5) + y = 1$.   6. Inconsistent. Hint: $(1 - y) + y = 2$. So, $1 = 2$. False.   7. (1, 2) One method: $y = -2x + 4$. Then $3x - (-2x + 4) = 1$ or $3x + 2x - 4 = 1$.   8. (3, -1)   9. Dependent. $2 = 2$.   10. (0, 1)

11. Inconsistent   12. (-7, -1)   13. (-3, -7)   14. (0, 0)   15. $(-\frac{1}{2}, -1)$ 16. Inconsistent

17. Dependent   18. check, $(\frac{3}{2}, -2)$ $2x - 3y = 9$ $6x + 2y = 5$

$2(\frac{3}{2}) - 3(-2) = 9$ $6(\frac{3}{2}) + 2(-2) = 5$

$3 + 6 = 9$ $9 - 4 = 5$

$9 = 9$ True. $5 = 5$ True.

95

Systems, Exercise 4, page 34.

1. (-1, 2)   2. (-3, -1)   3. (5, -2)   4. Inconsistent   5. (1, 1)   6. (2, 1)   7. (2, -2)   8. Dependent
9. (8, -11)   10. (1, -1)   11. (-1, 1)   12. check (2, - 5)
13. Multiply $2x - 3y = 9$ by -3, both sides, then add equations to get $11y = -22$. Proceed from there.

Rational Expressions, Exercise 1, page 37.

1. Terms.   2. Terms   3. Factors   4. Factors   5. Factors   6. Factors   7. Legal to cancel factors. Illegal
to cancel terms.   8. 3   9. $\frac{y}{z}$ 10. $\frac{3}{4}$ 11. $\frac{x+1}{5}$ 12. Not reducible   13. $\frac{1}{x}$ 4. $\frac{x}{y+7}$ 15. Not reducible
16. $\frac{3}{4}$ 17. $\frac{1}{3}$ 18. Not reducible   19. $(x + 2)$ Hint: Example 2 .
20. $\frac{y^2 + 4}{y + 2}$ 21. Not reducible   22. $\frac{1}{x}$ 23. $\frac{x+2}{x+3}$ 24. Not reducible   25. $x + 2$ Hint: Example 3   26. $\frac{x+3}{x+2}$
27. $\frac{y+3}{y-3}$ Hint: $y^2 - 9 = (y + 3)(y - 3)$ 28. Not reducible   29. $\frac{x+y}{x-y}$

Rational Expressions, Exercise 2, page 40.

1. $\frac{49}{12}$ 2. x   3. xy   4. $2(x + 3)$ 5. x - 2   6. $\frac{2}{15}$ 7. 1   8. $\frac{x}{x+1}$ 9. x - 2y Hint: $x^2 - 6xy + 8y^2 = (x - 2y)(x - 4y)$.

Rational Expressions, Exercise 3, page 42.

1. 90   2. 180   3. $x \cdot x \cdot y$ 4. $x(x + 1)(x + 2)$ 5. $x(x + 1)(x + 2)$ 6. $y(y - 3)(y + 4)$
7. $(y + 4)(y - 4)(y - 3)$ 8. $(x + 2y)(x + y)(x - y)$ 9. 4x   10. 9y   11. 35   12. xy
13. $x(x + 2)$ 14. $xy(x + 2)$ 15. xy

Rational Expressions, Exercise 4, page 45.

1, $\frac{2x + 4}{x}$ 2. 5 Hint: $5y + 5 = 5(y + 1)$ 3. $\frac{x + 10}{2(x - 3)}$ 4. $\frac{4}{y + 2}$ Hint: $3y + y = 4y$, then ys cancel.   5. $\frac{x + 6}{(x + 3)(x + 4)}$
6. $\frac{2}{x - 3}$ 7. $\frac{-2x + 10}{x}$ 8. 4   9. $\frac{1}{2}$ Hint: $2x - (x + 3) = x - 3$ which cancels.   10. $\frac{x^2 + x - 5}{x(x + 2)}$ 11. $\frac{2y + 6}{y(y - 2)}$ 12. $\frac{y - x}{xy}$

Rational Expressions, Exercise 5, page 48.

1. 5 LCM = 6   2. $\frac{7}{20}$ LCM = 12   3. y + x   4. $\frac{1 - 2y}{3x}$ LCM = xy   5. $\frac{x + 3}{x}$ 6. 5x + 21   7. $\frac{1 + xy - y^2}{x + y}$

Rational Expressions, Exercise 6, page 51.

1. 1   2. 1   3. 1   4. -3   5. -7   6. 2   7. No solution since x = 0 which makes the denominator 0.
8. 7 and 3   9. 2 and 3   10. No solution. You get 0 = -1, false.   11. - 3 makes denominators - 4, 12 and - 3.

Review, Chapters 1, 2, and 3, Page 52.

Chapter 1

1.

3 a.

2a. No   b. Yes   c. Yes

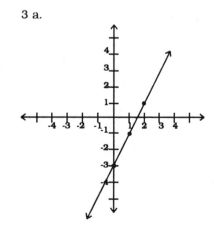

3 b.
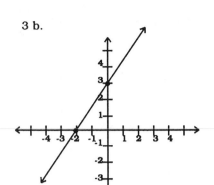

3 c.
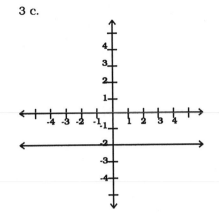

4. $-\dfrac{3}{4}$

5a. Slope = 9, y-intercept = 1

5b. Slope = $-\dfrac{1}{2}$, y-intercept = - 3

5c. Slope = -2, y-intercept = 3.

6. Same graph as in 3a. and 3b.

7. $y = -\dfrac{1}{2}x$. Hint: b = 0

Chapter 2

1a. No   1b. Yes   2a. Yes   2b. No   3a. (-2, -3)   3b. (-2, -3)   4. Dependent all methods.

Chapter 3

1a. $\dfrac{1}{y}$ 1b. $\dfrac{1}{x+3}$ 1c. x - 5   1d. $\dfrac{x-3}{x}$ 2a. $\dfrac{5}{x+1}$ 2b. $\dfrac{x+2}{x+1}$

3a. $\dfrac{5}{x}$ 3b. $\dfrac{x}{x^2-4}$ 3c. $\dfrac{y-x}{xy}$ 4a. y + x   4b. $\dfrac{7}{x-5}$ 5a. 21   5b. 1

Radicals, Exercise 1, page 56.

1. Radical sign; radicand.   2. xy   3a. 7   3b. 8   3c. 1   3d. No such number.   3e. 0   3f. 2
3g. No such number   3h. $\dfrac{1}{5}$ 4a. Principal root = 9; negative root = -9   4b. Principal root = 10;
negative root = -10.   4c. Has no square root.   4d. 0   5a. 3   5b. -3   5c. No meaning.
5d. No meaning.   5e. 4   5f. -4   5g. -1   5h. No meaning.   5i. 1   5j. -10   5k. 0   5l. 6

Radicals, Exercise 2, page 57.

1. 49, 64, 81, 100, 121, 144, 169.   2a. 5   2b. 5   2c. 9   2d. c   2e. c   2f. 127   2g. 0.75
2h. x   2i. x + 3   2j. xy   2k. y + 2   2l. 2x + 7   3. Perfect square.

Radicals, Exercise 3, page 60.

1a.   Left side   Right side   1b. 10 both sides
$\sqrt{4\cdot16}$ $\sqrt{4}\cdot\sqrt{16}$
$=\sqrt{64}=8$ $2\cdot4=8$
2a. $\sqrt{3\cdot4}=\sqrt{3}\cdot\sqrt{4}$ 2b. $\sqrt{x\cdot x}=\sqrt{x}\cdot\sqrt{x}$ 2c. $\sqrt{4\cdot11}$ 2d. $\sqrt{x^3\cdot x}$
3a. $2\sqrt{3}$ 3b. $3\sqrt{2}$ 3c. $5\sqrt{2}$ 3d. $2\sqrt{5}$ 3e. $3\sqrt{5}$ 3f. $2\sqrt{2}$
3g. $3\sqrt{7}$ 3h. $4\sqrt{2}$ 3i. $4\sqrt{3}$ 3j. $6\sqrt{2}$ 3k. $10\sqrt{6}$ 3l. $6\sqrt{3}$
4a. $2\sqrt{2}$ 4b. $2\sqrt{7}$ 4c. $5\sqrt{3}$ 4d. $2\sqrt{10}$ 4e. $3\sqrt{3}$ 4f. $2\sqrt{6}$
4g. $3\sqrt{6}$ 4h. $4\sqrt{2}$ 4i. $2\sqrt{15}$ 4j. $10\sqrt{2}$ 4k. $6\sqrt{2}$ 4l. $10\sqrt{6}$

Radicals, Exercise 4, page 63.

1a. x^4 1b. y^3 1c. x   1d. 2^5 1e. x^3y^4 1f. 2^5y
2a. x^3 2b. 2^5 2c. y^{50} 2d. ab^4 2e. $y^3\sqrt{y}$ 2f. $x^5\sqrt{x}$ 2g. $y^2\sqrt{3y}$ 2h. $x\sqrt{x}$
3a. $4x^2\sqrt{x}$ 3b. $3x^2y\sqrt{3}$ 3c. $5b^3\sqrt{2ab}$ 3d. $10a^5b\sqrt{2a}$ 3e. $3x^2y^2\sqrt{11}$
3f. $4\sqrt{2xy}$ 3g. $5a^{10}bc\sqrt{3a}$ 3h. $4y^{12}\sqrt{7x}$ 3i. $5x^3z^6\sqrt{15xyz}$

Radicals, Exercise 5, page 65.

1. $2\sqrt{3}$ 2. $3\sqrt{2x}$ 3. 5 4. y^4 5. x 6. $2a^2\sqrt{2b}$ 7. $2\cdot2\sqrt{15}=4\sqrt{15}$ 8. $35x^2\sqrt{6}$ 9. $7\sqrt{6}$
10. $3x\sqrt{35}$ 11. $5y^3\sqrt{21}$ 12. $11x\sqrt{35y}$ 13. 14 14. $3a^5b^5\sqrt{11ab}$ 15. $2x^2$ 16. $5a^3\sqrt{a}$ 17. 20

Radicals, Exercise 6, page 66.

1. $2\sqrt{y}$ 2. $9\sqrt{2}$ 3. $6\sqrt{3}$ 4. $3\sqrt{y}+\sqrt{5}$ 5. $7\sqrt{2}$ 6. $-\sqrt{7}$ 7. $-3\sqrt{2}+2\sqrt{3}$ 8. $2\sqrt{x}$ 9. $11\sqrt{2}$
10. $\sqrt{2xy}$ 11. $5a$ 12. $8y^2\sqrt{y}$

Radicals, Exercise 7, page 70.

1. Left side: $\sqrt{\frac{36}{4}}=\sqrt{\frac{9}{1}}=\sqrt{9}=3$, Right Side: $\frac{\sqrt{36}}{\sqrt{4}}=\frac{6}{2}=\frac{3}{1}=3$

2a. 3 2b. $\sqrt{5}$ 2c. $\frac{\sqrt{x}}{2}$ 2d. $x\sqrt{5}$ 2e. $\frac{\sqrt{7}}{3}$ 2f. $\frac{5y^3}{2}$ 2g. $\frac{\sqrt{2}}{3}$

3a. $\frac{4\sqrt{3}}{3}$ 3b. $\frac{\sqrt{10}}{5}$ 3c. $\frac{\sqrt{3x}}{x}$ 3d. $\frac{5\sqrt{2}}{14}$ 3e. $\frac{\sqrt{6}}{15}$

3f. $\frac{2\sqrt{2}}{2}=\sqrt{2}$ 3g. $\sqrt{x}$ 4a. $\frac{5}{\sqrt{2}}\cdot\frac{\sqrt{2}}{\sqrt{2}}=\frac{5\sqrt{2}}{2}$ 4b. $\frac{3\sqrt{5}}{5}$ 4c. $\frac{\sqrt{6}}{3}$ 4d. $\frac{\sqrt{3x}}{3}$

4e. $\frac{\sqrt{xy}}{y}$ 4f. $\frac{\sqrt{2a}}{2}$ 4g. $\frac{\sqrt{2x}}{x}$ 4h. $\frac{\sqrt{x}}{2x}$

Radical Expressions, Exercise 8, page 73.

1. $x=100$ 2. 16 3. 25 4. 27 5. 2 6. 2 and 1 7. 10 only, since 5 does not check. 8. 5
9. 1 10. No solution. Answer does not check: $\sqrt{4}=2$, not -2.

Quadratic Equations, Exercise 1, page 76.

1. ±8 2. $\pm2\sqrt{3}$ 3. $\pm5\sqrt{2}$ 4. ±4 5. ±2 6. ±1 7. -7 or 1 Hint: Example 3. 8. 4 or -2
9. $\frac{4}{3}$ or 0 10. $\frac{-1\pm2\sqrt{3}}{2}$ Hint: Example 4. 11. $\frac{5\pm3\sqrt{2}}{3}$ 12. No solution, $\sqrt{-4}$ has no meaning.
13. $\sqrt{2}=1.4142$. So $\frac{1+3\sqrt{2}}{5}=1.048$. $\frac{1-3\sqrt{2}}{5}=-.6485$.
14. $2(2\sqrt{3})^2-9=15$
$\quad\quad2(4\cdot3)-9=15$
$\quad\quad\quad2(12)-9=15$
$\quad\quad\quad\quad\quad15=15$

Quadratic Equations, Exercise 2, page 77.

1. 0 and -5 2. 0, 1 3. 0, $\frac{5}{7}$ 4. 5, -5 5. 2, -2 6. $\frac{7}{2}$, $-\frac{7}{2}$ Hint: $(2x+7)(2x-7)=0$ 7. 4, -4
8. 7, -3 9. -1 10. 2, $-\frac{3}{2}$

Quadratic Equations, Exercise 3, page 81.

1. $a=1$, $b=2$, $c=-3$. Answer $x=1$ or -3. 2. $\frac{1}{2}$ or 1 3. -3 4. No solution
5. 0 or 1 6. $\frac{1}{2}$ .First add 1 to both sides. 7. No solution
8. $\frac{7\pm\sqrt{41}}{2}$ 9. $\pm\sqrt{5}$ 10. No solution 11. $\frac{-4\pm\sqrt{20}}{2}$ 12. $b^2-4ac=0$. One solution,
13. No solution 14. Two solutions 15. No solution 16. Two solutions. Hint: Subtract 5x first.
17. One solution 18. 4, two solutions 19. 0, one solution 20. -19, no solution
21. $x=-1$

Quadratic Equations, Exercise 4, page 82.

1. 4 and 0 by the square root method.   2. 1 and 2 by the zero product method.

3. $\dfrac{-5 \pm \sqrt{17}}{2}$ by the quadratic formula.   4. 5 or -5 by all three methods.   5. $\dfrac{1}{2}$ and $-\dfrac{2}{3}$ both methods.

Quadratic Equations, Exercise 5, page 85.

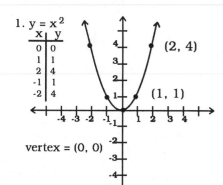

1. $y = x^2$

x	y
0	0
1	1
2	4
-1	1
-2	4

(2, 4)

(1, 1)

vertex = (0, 0)

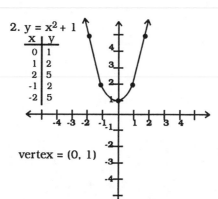

2. $y = x^2 + 1$

x	y
0	1
1	2
2	5
-1	2
-2	5

vertex = (0, 1)

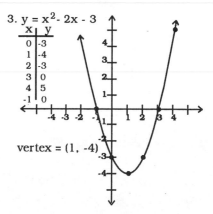

3. $y = x^2 - 2x - 3$

x	y
0	-3
1	-4
2	-3
3	0
4	5
-1	0

vertex = (1, -4)

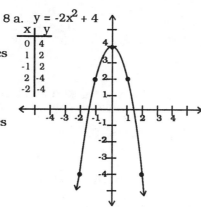

8 a. $y = -2x^2 + 4$

x	y
0	4
1	2
-1	2
2	-4
-2	-4

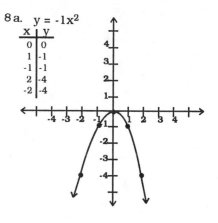

8 a. $y = -1x^2$

x	y
0	0
1	-1
-1	-1
2	-4
-2	-4

4. $b^2 - 4ac = 16$. A positive discriminant indicates two x-intercepts. Thus, the discriminant agrees with the graph of Exercise 3.

5. x = 3 or -1. Yes, these are identical to the x-intercepts on the graph.

6. $b^2 - 4ac = -4$. A negative discriminant indicates there are no x-intercepts and the parabola of Exercise 2 agrees with this.

7. $b^2 - 4ac = 0$, which indicates one x-intercept. The graph of Exercise 1 agrees with this.

Review, Chapters 4 and 5, page 86.

Chapter 4

1a. 8   1b. -3   1c. $3\sqrt{3}$ 1d. $6\sqrt{2}$ 1e. x^3 1f. $2y^2 \sqrt{3}$ 1g. $x^2 \sqrt{x}$ 1h. $y^3 \sqrt{3y}$
1i. $3x^5 \sqrt{2y}$ 1j. 6   1k. x   1l. $10\sqrt{5}$ 1m. $14\sqrt{2}$ Hint: prime numbers.
2a. $2\sqrt{5}$ 2b. 7   2c. $3\sqrt{5x}$ 2d. x   2e. y^2 2f. $4xy\sqrt{3}$

3a. $12\sqrt{x}$ 3b. $4\sqrt{2}$ 3c. $-1\sqrt{3x}$ 4a. $\dfrac{2}{3}$ 4b. $\dfrac{3}{4}$ 4c. $\dfrac{x}{2}$ 4d. $\sqrt{2}$ 4e. $\dfrac{5\sqrt{2}}{2}$

4f. $\dfrac{\sqrt{14}}{2}$ 4g. $\dfrac{10\sqrt{x}}{x}$ 4h. $\sqrt{3}$ 4i. $\dfrac{\sqrt{2x}}{2}$ 4j. $5\sqrt{y}$ 5a. 25   5b. 36   5c. -2

Chapter 5

1a. $\pm 2\sqrt{2}$ 1b. ± 1 1c. 0 and 6. Hint: $x = 3 \pm 3$ 2a. 0,8   2b. 2, -2   2c. 3, 4
2d. $-\dfrac{1}{3}, 2$ 2e. $\dfrac{-3 \pm \sqrt{5}}{2}$ 2f. No solution   2g. $\dfrac{-1 \pm \sqrt{5}}{2}$ 2h. 1 and $\dfrac{1}{2}$ 2i. $-\dfrac{5}{3}, -\dfrac{2}{7}$
3a. 21. Two solutions.   3b. -35. No solution.   3c. 0. One solution.
3d. 0. One solution.   3e. 64. Two solutions.   3f. 25. Two solutions.

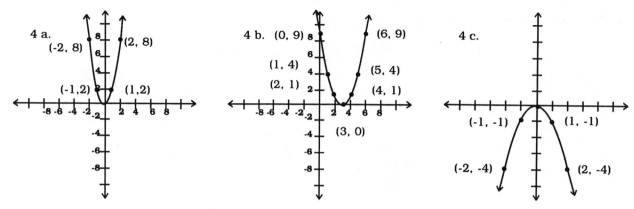

4 a. (-2, 8) (2, 8) (-1,2) (1,2)

4 b. (0, 9) (6, 9) (1, 4) (5, 4) (2, 1) (4, 1) (3, 0)

4 c. (-1, -1) (1, -1) (-2, -4) (2, -4)

5. $\frac{1}{2}$ and 4. Hint: replace y by 0.   Use the quadratic formula

Cumulative Review, page 88.

1a. slope = -3
     y-intercept = 7

1b. slope = $\frac{4}{5}$
     y-intercept = -6

1c. slope = -1
     y-intercept = 0

1d. y = 5x + 1
    slope = 5
    y-intercept = 1

1e. y = 3x
    slope = 3
    y-intercept = 0

1f. y = 2x + 16
    slope = 2
    y-intercept = 16

2.

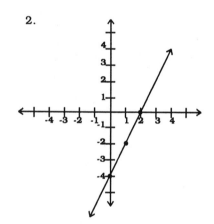

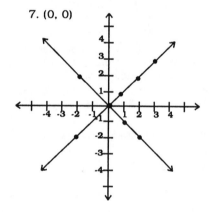

3. Same graph as Exercise 2.
4. Same graph as Exercise 2.
5. 1
6. y = -3x + 1

7. (0, 0)

8. (0, 0)
9. (0, 0)
10. (1, -3) each method.
11. Inconsistent.
12. Dependent.

13a. $\frac{1}{y}$ 13b. $\frac{x+2}{2}$ 13c. $\frac{1}{5}$ 13d. x + 9   13e. $\frac{1}{x}$ 13 f. $\frac{1}{x+6}$

14a. x   14b. $\frac{1}{x+5}$ 14c. x + 2   14d. $\frac{3y}{y+1}$ 15a. $\frac{x+2}{2x}$ 15b. $\frac{1}{x}$

15c. $\frac{2x+1}{(x+1)(x+1)}$ 16. x + 2   17. -1

18a. 10   18b. 0   18c. $2\sqrt{2}$ 18d. $2\sqrt{5}$ 19a. $6\sqrt{15}$ 19b. p   19c. $7x\sqrt{6}$

20a. $\sqrt{2}$ 20b. $11\sqrt{x}$ 21a. 2   21b. $\sqrt{5}$ 21c. $\sqrt{7}$ 21d. $\frac{\sqrt{5x}}{x}$ 22. 4

23a. ± 10   23b. 2 or -1   23c. -2 ± $\sqrt{10}$ 24a. 0, 7   24b. - $\frac{2}{3}$, $\frac{1}{2}$ 24c. 7, -7

25a. $\frac{5 \pm \sqrt{13}}{2}$ 25b. ± $\frac{\sqrt{12}}{6}$ = ± $\frac{\sqrt{3}}{3}$ 25c. No solution

26a. ± $5\sqrt{2}$ 26b. -2, -9   26c. $\frac{-3 \pm \sqrt{5}}{2}$ 26d. $\frac{3}{2}$, - $\frac{3}{2}$ 26e. No solution   26f. ±4

100

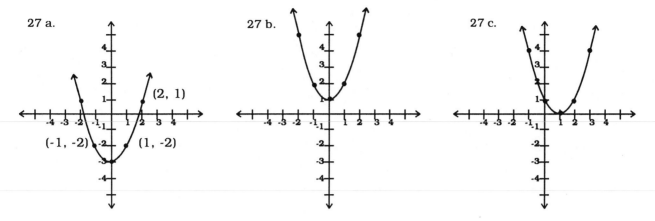

27 a.

(2, 1)

(-1, -2)   (1, -2)

27 b.

27 c.

28a. $b^2 - 4ac = 12$. Two x-intercepts.   28b. - 4. No x-intercept.   28c. 0. One x-intercept.

English Series

The Straight Forward English Series is designed to measure, teach, review, and master specific English skills: capitalization and punctuation; nouns and pronouns; verbs; adjectives and adverbs; prepositions, conjunctions and interjections; sentences; clauses and phrases; and mechanics.

Each workbook is a simple, straightforward approach to learning English skills. Skills are keyed to major school textbook adoptions.

Pages are reproducible.

GP-032 Capitalization and Punctuation
GP-033 Nouns and Pronouns
GP-034 Verbs
GP-035 Adjectives and Adverbs
GP-041 Sentences
GP-O43 Prepositions, Conjunctions, & Interjections
ADVANCED SERIES, large editions
GP-055 Clauses & Phrases
GP-056 Mechanics

Substitute Teaching

GP-027 Substitute Teacher Folder
A pertinent information folder left by regular classroom teachers listing class schedules, classroom procedures, discipline, support personnel, and regular classroom teacher expectations.

GP-001 Substitute Ingredients
A collection of imaginative language arts, math, and art activities for grades 3–8. Reproducible master sheets accompany most lessons.

GP-002 Mastering the Art of Substitute Teaching
Substitute teaching formats, strategies, and activities strictly from practical experience.

GP-003 Classroom Management for Substitute Teachers
Suggested procedures for being-in-charge, establishing rapport, and getting the support of regular classroom teachers and staff.

GP-014 Lesson Plans for Substitute Teachers
A packet of 12 lesson plan forms to be filled out by regular classroom teachers to provide one day of instruction during their absence.

GP-004 Just Fun
Engaging, high-interest activities that are short span, 10-15 minutes in length.